T0291604

The Cambridge Nature Study Series

General Editor : HUGH RICHARDSON, M.A.

WEEDS

Plate I

Invasion by Weeds

This place was a potato-patch five years previously. It now supports a dense population of weeds, mostly native plants. Ox-eyes and Purple Loosestrife predominate, with Wild Angelica (in front on left), Field Thistle (behind on left, Sow-Thistle (in front of figure), Marsh Thistle (on right of figure) and other species. Photographed by R. Welch near Cavan, Ireland.

WEEDS

SIMPLE LESSONS FOR CHILDREN

BY

ROBERT LLOYD PRAEGER

WITH ILLUSTRATIONS BY

S. ROSAMOND PRAEGER AND R. J. WELCH

Cambridge :

at the University Press

1913

CAMBRIDGE
UNIVERSITY PRESS

University Printing House, Cambridge CB2 8BS, United Kingdom

Cambridge University Press is part of the University of Cambridge.

It furthers the University's mission by disseminating knowledge in the pursuit of education, learning and research at the highest international levels of excellence.

www.cambridge.org
Information on this title: www.cambridge.org/9781316613184

First published 1913
First paperback edition 2016

A catalogue record for this publication is available from the British Library

ISBN 978-1-316-61318-4 Paperback

PREFACE

THE question of weeds, and how they can be kept in check, is one of the most important which has to be faced by the farmer or gardener. Agriculture is an eternal war against weeds, and by neglect of cleaning his land, the careless cultivator may lose one-fourth or in extreme cases even one-half of the yield which might be his. The preaching of the crusade against weeds cannot be begun too early. In order to fight weeds, we must know how they grow and how they spread; to these points particular attention is given in the following chapters.

Apart from their importance as agricultural pests, weeds are, as such, a fascinating study. The history of weeds is the history of agriculture. Ever since the peoples of the Stone Age first began to till the ground, weeds and crops have gone hand in hand; and man's migratory movements, his wars, and later on his trade, have all played their part in assisting the spread of his enemies the weeds.

Again, our weeds include many beautiful and interesting plants; this study is most instructive; the structure of flowers, the dispersal of seeds, the function of the different parts of plants, and indeed all the lessons of elementary botany, can be studied to full advantage among our common weeds. In the pages which follow I have attempted to make clear the great interest of weeds, and the romance which attaches to them, just as much as to set forth the serious injury which they inflict when they are not kept within bounds.

While these lessons will be found applicable to any portion of the British Isles, I have preferred to draw particular illustrations from the area with which I am most familiar; hence the frequent mention of Ireland in the following pages.

Within the limits of so small a book, it has not been possible to deal with the practical side of the question of weeds—their enumeration, description, the methods used or suggested for their extermination, and the principles of seed-testing—save by a few examples and suggestions; but the book is intended for the use of school children, not of farmers. For the latter, Long's excellent work[1] is available; and Pammel's treatise[2], though dealing with American agriculture, is full of suggestion for the European cultivator.

To my friend Prof. James Wilson, M.A., my thanks are due for some useful suggestions.

R. Ll. P.

Dublin,
August, 1913.

---

[1] Harold C. Long : *Common Weeds of the Farm and Garden.* 8vo. London : Smith, Elder & Co. 1910. Price 6/-.

[2] L. H. Pammel : *Weeds of the Farm and Garden.* 8vo. New York and London (Kegan Paul). 1911. Price 7/6.

CONTENTS

CHAPTER I

What Weeds are, and their place in the Plant World

CHAPTER II

The Life of a Plant

CHAPTER III

On Weeds in General

CHAPTER IV

Seeds and their Ways

CHAPTER V

The War against Weeds

CHAPTER VI

Some Common Weeds

ILLUSTRATIONS IN THE TEXT

Fig. 12 is modified from Sowerby's *English Botany*, Vol. ix, Plate MDXXIV, by permission of Messrs G. Bell & Sons, Ltd.; fig. 43 (right-hand drawing) is based upon Kerner's *Natural History of Plants*, Vol. i, p. 439, by permission of Messrs Blackie & Son, Ltd.

PLATES

CHAPTER I

WHAT WEEDS ARE, AND THEIR PLACE IN THE PLANT WORLD

WHEN there were no men, there were no weeds. This statement requires explanation, and the explanation involves the question of what we mean by weeds. To understand this question in all its bearings, we must in the first place consider briefly the general subject of plant-life upon our earth.

The plant-world. One of the most remarkable features of this globe upon which we live is the mantle of vegetation with which the greater part of the land-surface (as well as the shallower waters) is clothed. This vegetation has been evolved during a vast period of time. It has assumed an endless variety of forms—from the towering forest tree down to the microscopic diatom—and it has accommodated itself to a wide range of physical conditions. In varying form, plant-life is found in the intense cold of the Arctic Regions and the fierce heat of the Equator; some plants grow under thirty feet of fresh or salt water, others in deserts where only a few showers fall; and while some species endure the full glare of the tropic sun, others can live in per-petual darkness. But taking the green-leaved flowering plants which we all know so well, and with which we shall be mainly concerned in the pages which follow, we find that there are physical limits to their existence. A certain amount of heat and light, and air and water they must have ; and also they need to have access, by means of their roots, to certain mineral substances which they find in the soil. The absence of one or more of these requirements produces a desert—a place where plants are absent. The want of heat produces a dwindling of the vegetation of the far north ; want of water is accountable for the scanty flora of the Sahara, and the absence there of a covering of plants in its turn allows the surface to become disintegrated,

Fig. 1.   A Stonecrop (*Sedum album*), showing thickened leaves for the storing up of water.   Natural size.

so that the loose soil is tossed to and fro by every wind. Want of light is the cause of the dying out of vegetation in the deeper parts of lakes, and of sea-weeds in the oceans.   Want of soil in which the plants can anchor themselves by means of their roots, and from which they can draw their mineral food, accounts for their absence on certain rocky tracts.   The too great abundance in the soil of certain common mineral substances, such as common salt or compounds of potash, which in minute quantity are useful to plants, or at least not harmful to them, also cause a dearth of vegetation, as in the alkali deserts of the United States.

But inside of these extremes, plants usually abound, and cover the earth with their verdure. By slow degrees, too, they have accommodated themselves to special conditions of existence. Plants which grow in very dry places protect themselves from too great loss of water by having small leaves, or often no leaves at all; by having a thick waterproof skin, or by covering themselves with a layer of wax or of dense hairs to keep themselves cool and moist; often by thickening their leaves and stems, and storing up quite a large quantity of water there, to be used in time of drought (Fig. 1). Plants exposed to very strong sunlight often hang their leaves edge-wise, like the gum-trees of Australia, to prevent their being burned up. Plants which grow in shady places on the other hand have mostly large spreading leaves, so as to catch plenty of light. In very cold places plants protect themselves in various ways. Water-plants have limp stems like whipcord, which are less liable to get broken by waves than if they were stiff, and their leaves, which have not to resist wind or rain or hail, are mostly very thin.

The struggle for existence. The number of seeds which plants produce is far greater than the number of seedlings which survive. There are many species which bear so much seed that if every seedling grew, they would in quite a small number of generations cover the whole land-surface of the globe. The effect of the incessant crowding of both young and old plants which is always going on, is that the weaker ones are killed off, while the stronger ones—in other words, the ones which are best fitted for the particular situation in which they grow—tend to survive. Thus any slight variation which helps a plant to hold its own will tend to continue, because the plant which possesses it will be more likely to flourish and to produce seed than its neighbours, and the children tend to have the same characters as the parent. This is the simple principle which underlies the theory of Natural Selection, which will be always associated with the name of the great naturalist, Charles Darwin, who first propounded it in 1859.

If we take, then, any area of natural vegetation, we must realize that the different kinds of plants which we find within it are not there by chance. Many kinds of plants from all around are trying to seed themselves there or to creep in, and only those which are in every way best suited for the particular conditions which prevail in any one spot will be able to maintain their hold. In every kind of situation—mountain, river, lake, wood, bog, or sandy shore—the plants which naturally grow there have a long history of struggle and adaptation behind them, and each maintains its place in virtue of its special capacity to do so.

Enter man. Into this beautifully balanced plant-world, many thousands of years ago, entered man. So long as he lived in the forests or on the plains, hunting, fishing and eating wild fruits and roots, he did not affect the natural vegetation as much as did the herds of wild grazing animals which were his prey. When he tamed some of these grazing animals, such as horses, cows, and sheep, and moved about on the natural grass-lands, driving his herds before him, he still affected the vegetation only to a slight degree. But when he learned the art of tillage, and began to dig up the grass-lands, and with the aid of fire to clear away the forests, in order that he might grow in their place plants useful to himself, then he attempted to substitute for the old native plant-covering an entirely artificial vegetation, and then he commenced a struggle with the natural flora which has been going on ever since, and still goes on continuously. The native plants are always striving to regain possession of their former home, and man works incessantly to keep them out; these natural invaders of man's territory form a large section of the great group of plants undesired by man which are known to him as **weeds**. Weeds of this kind—old natives which are ever trying to regain their lost empire —are undoubtedly the kind of weeds with which the ancient races of mankind had to deal first, and they are still the most abundant kind in districts which are not wholly reclaimed from a state of nature, or where agriculture is in a backward condition. In the wilder parts of the west of Ireland, for instance, most of the weeds in the farm-land are plants which originally occupied the ground, and which still grow as natives close by; by means of their seeds, or their creeping stems, they are incessantly invading the tilled land, and the farmer is constantly occupied in keeping them out, by weeding, or digging, or ploughing.

There is another class of weeds in our fields, which has been brought there—though unintentionally and unwillingly—by man himself, and which differs widely from the former group inasmuch as it is maintained there *because of*, and not, like the former group, *in spite of* man's operations. To understand this group fully, we must consider briefly the various types of vegetation which occur naturally in Europe, and the peculiarities of climate which produce them.

The vegetation of Europe.   All around the basin of the Mediterranean there is a prevalence of dry north-east winds, blowing from the adjoining masses of land, during the summer months, while in winter wet south-westerly winds, blowing in from the ocean, are frequent. The result is a hot dry summer, and a wet mild winter. The drought and heat of summer appear to be the controlling factors as regards vegetation. The annual plants hurry through their life during the spring months, and are in seed by the time the great heat arrives ; the perennials either hurry likewise, and lie dormant during the summer ; or if they face the heat, they do so with the aid of the devices which have already been mentioned —small leaves, succulent leaves, long roots, and so on. Annuals are abundant, and tall forest trees are not characteristic.   When we go northward of this region, we find the moisture-laden westerly winds prevalent, giving a good supply of rain all the year round.   This allows forests to grow, for forests, on account of the great amount of water which the trees use, need a more abundant rainfall than bushes or small-leaved herbs ; and in spite of all man's interference with the natural vegetation, forests are still a characteristic feature of middle Europe, often with grass-lands on the hills.   To the north of this again, the increasing cold of higher latitudes limits the spread of the forests, and they again degenerate into scrub, and pass into creeping shrubs and grasses.   Now, the horde of annual plants of the Mediterranean region was not fitted to compete with the dense population of perennial plants—trees, herbs, and grasses—which occupied the lands to the northward, so long as this dense covering of vegetation remained intact.   But when man began to destroy the forests and dig up the ground on which they had grown, in order to cultivate grains, then annual plants—both those native to the vicinity and those from the region further south —gained an advantage. This was both because the open ground was more suitable to their habits, and because of their rapid growth, which allowed them to spring up and sow themselves again before the next digging or ploughing rooted them out. And in this way the spread of tillage from the old countries of the Mediterranean northwards and westwards across the former forest-lands was accompanied by a spread of many of the annual plants of the Mediterranean region. Furthermore, much of the seed which was sown in the fields came—as it still does—from the Mediterranean countries, and mixed with this seed were the seeds of many of the annual weeds of the fields there. In this manner man himself helped, and still helps, the spreading of a large number of weeds. As we said at the beginning of this section, man—though unintentionally and un-willingly—himself introduces into his fields many of the most unwelcome weeds.

What weeds are. We are now in a position to explain the statement with which this chapter opened— that when there were no men there were no weeds; and also to define what is meant by weeds. Before man began to interfere with the natural vegetation, plants grew only where they were best fitted to grow ; each species and group of species occupied a definite place in the plant societies, and maintained itself there so long as existing conditions remained ; and speaking generally, conditions changed only very slowly—owing to alterations in drainage, for instance, such as the silting up of lakes, or the cutting down by streams of their channels; or the much slower but much greater changes brought about by alteration of climate.

Man, in his agricultural operations, entirely reverses this order of things. He exterminates, or tries to exterminate, the natural vegetation, and he cultivates instead plants which are specially useful to himself, but which are not natural to the vicinity and which can exist there only under his protection. As has been said already, the native plants all the time strive to break in and re-occupy their old homes, and men are as constantly employed in driving them back; thus agriculture has been defined as "a controversy with weeds." **Weeds are, in fact, plants growing in places where man does not want them to grow.**

It may be well, at this stage, to refer to a use of the word "weed" which, though commonly used, is not correct. It arises in this way. The plants which man grows are selected for either their use or their beauty. Most of the true weeds which creep in among them, and which man employs himself in combating, are plants of no special use and not particularly showy. Hence the term "weed" comes to be applied to any plant which is not directly useful to man, nor ornamental, even though it may be a native species growing in quite natural surroundings, and not affecting man and his operations in any way. This is not the sense in which the term "weed" is used in the following pages.

The vegetation of Ireland. Ireland, to which frequent reference will be made for illustration, lies within the meteorological influences which have produced the forest region of Europe. Indeed, on account of its position on the eastern edge of the North Atlantic, it has a climate which is milder and wetter than that of any other country in Europe. The heavy rainfall has tended to produce not only forests (of which traces only remain now-a-days), but also the great peat-bogs which are so familiar a feature of the Irish landscape. The forests vanished during the last thousand years before the axe and plough of man and his herds of cattle. In some districts, where metallic ores occurred, they were used up largely for the purpose of smelting; no doubt they also long supplied the domestic hearth. The needs of agriculture accounted for the disappearance of the remainder. At the present time we see the plains and hill-slopes laid out in square fields with hedges between; as we go westward, where on account of the high winds trees grow with difficulty, fences of earth or stone replace the hedges of the east. The old forests have passed away. Only here and there, in steep glens, and on ground too rocky or too poor for cultivation, we find remnants of the primitive woodland, and from these remnants we may learn still what the native trees were that formerly occupied so much of the country. Oak was abundant, and with it grew Ash, Elm and Birch, Poplar and Alder, and other smaller trees, such as Hazel, Holly, Mountain Ash, Yew, Hawthorn, and so on. Many of the trees most familiar to us now, such as Sycamore, Horse Chestnut, Beech, Larch, Spruce, were brought into Ireland from other countries by man.

The other great natural plant-formation of Ireland, the bogs, has proved more difficult to deal with than the forests, and much of it remains still. On the forest-land, once the trees were cleared away a rich soil lay

ready for use. But the bog surface, even when thoroughly
drained and dug, is not suitable for the growth of most
plants, being charged with injurious substances (humous
compounds) derived from the incomplete decay of the
plants which form the bog. Were it not for the fact
that turf is a valuable fuel, the Irish bog-lands would
still extend almost untouched by the hand of man. But
during many centuries great areas of bog have been
removed and burned, and on the lower layers of the bog
which are thus reached, and which are more decayed
than the upper layers, crops can be grown, especially
when plenty of lime and manure are added to the peaty
soil, and when it is sufficiently drained. Near the
bottom of the bogs are found, often in great quantity,
stumps of the Scotch Fir. In early times this tree and
the plants of the bog appear to have struggled for
mastery through a long period. Under dry conditions
the Fir spread over the country; when the ground
became wetter, it died out and the bog plants took
its place. This story is told by the bogs themselves,
in which we may often see more than one layer of old
Fir stumps, with layers of bog in between. What it was
that occasioned these fluctuations of vegetation it is not
easy to say now. In the end, the bog flora triumphed,
and over the top of the old tree-stumps it has in many
places built up as much as twenty feet of peat. But
at present, owing seemingly to a slight change of climate,
the building up of the bogs has in most places ceased,
and they are stationary, or in some places, as on the tops
and slopes of mountains, are being stripped off by rain
and wind. As to the Scotch Fir, unlike the other trees,
such as Oak, which often accompany it in the bog
deposits, it has as a native entirely died out in this

Plate II

Peat and Ancient Forests

View near Sligo, Ireland. Ten or twelve feet of peat have been removed for fuel, laying bare the remains of great forests of Scotch Fir which occupied the ground between the earlier stages of the peat growth. A few feet of peat lie below the layer of tree stumps. In the middle distance the old bog-surface has been cultivated, and has yielded a crop of hay. In the background are high hills of Carboniferous limestone. Photographed by R. Welch.

country, and there is every reason to believe that its disappearance, or at least its main decrease, took place at an earlier period than that of the forests which man destroyed, and that it was due to some change of climate. The Scotch Firs that are now abundant in Ireland were all planted by man, having been imported from other countries. But it is interesting to note that the conditions under which the Scotch Fir flourished in old days appear to have returned to some extent now, for self-sown seedlings grow readily, and a fir-wood is inclined to extend its boundaries just like a wood of one of our native trees.

In some places, as for instance on the slopes of mountains which were too exposed for trees to grow, and too dry for bog to form, grasses long ago occupied the ground. In our climate, in those places where the exposure is too great for trees to grow, it is also too great for crops to grow. Thus, these grassy lands have never been ploughed up, and they still remain almost unchanged.

Thus in Ireland a great part of the surface consists of old forest-land from which the trees have long since been cleared away, and which now presents the appearance of orderly fields separated by fences, dotted with cottages, and traversed by roads and lanes; the greater part of the ground is now pasture-land, the rest being under crops, mainly potatoes, oats and turnips. Next in importance comes the bog-land, much of which remains still; where it has been cut away for fuel, it is replaced by flat peaty fields, where potatoes and oats are grown with fair success. In addition, natural grass-land sometimes occurs. The marshes and swamps, which in old times covered wide areas, especially along the banks of rivers, have to a large extent been drained and converted into farm-land.

In the chapters which follow, it will be our duty to study the weeds which are found in these different kinds of ground—what they are, how they grow and spread, and how the farmer endeavours to keep them down.

EXERCISES I.

1. Bring indoors half a spade full of garden soil. Spread it out in a shallow box or pan. Cover with glass and keep it warm and damp. Pull up and count the seedlings which sprout in it.

2. Mark out a square yard of garden soil. Weed and rake the surface. Pull up and count seedlings as they appear.

3. Collect specimens of all the annual plants you can find growing on a fallow field. Write the names of those you know. Draw or describe those you cannot name.

4. Examine a neglected pasture field. Certain plants will be found standing uneaten among the grass. Collect specimens of these. Name, describe or draw.

5. Examine the ground where a wood has been felled within the last few years. Can you find any newly arrived annuals, biennials, or perennials beginning to grow there ?

6. Examine any ground where Gorse, Heather or Woodland has recently been burnt. Describe the new flora. Have the old plants survived, or have new plants arisen from seed ?

7. Describe the present flora of any land which has recently been drained or which is shortly to be drained.

8. Describe any remnants of primitive woodland which you can find among mountain crags or on islands in lakes or by swampy river sides.

9. Can you find any self-sown trees growing on peat soils ?

10. Describe what you see at the bottom of any bog from which all the peat has been cut away.

11. Examine with a magnifying glass a sample of agricultural seeds—clover, turnip or grass. Pick out any seeds which differ from the rest of the sample.

CHAPTER II

THE LIFE OF A PLANT

BEFORE we go deeper into the question of weeds, and try to understand where they come from and in what way they come, how they grow and seed and spread, and by what means their increase may be checked, it is necessary that we should have clear ideas as to the life-history of plants in general, and that we should obtain a grasp of the elementary principles which underlie both their internal economy and their external relations.

Life. The plant-world and the animal-world together differ from the other material things on our globe, such as the soil, the rocks, the water, or the air, by possessing that mysterious active quality which we call **life.** Plants and animals are built up of the same simple substances, which we call elements—such as oxygen, hydrogen, nitrogen, and carbon—which go to build up the inorganic world to which the air and water and rocks belong ; but they differ from them especially in their power of growth, and also of spontaneous movement, which is so familiar a feature of animal life, and is possessed in a less striking degree by plants. But one of the most remarkable, although most familiar conditions of organic growth is that it cannot go on indefinitely. Sooner or later the vital activities slacken and then cease—plants and animals grow old and die—growth and movement are suspended, and the plant-bodies and animal-bodies are dissolved again into the lifeless material of the universe out of which new life is built up. But life on the earth does not cease on account of the death of individuals. The activities of every plant and animal are concentrated on two main objects. The one is the building up of its own body and the carrying on of its own life. The other is the perpetuation of its kind by the production of fresh individuals. In the case of many of the lower forms of life, this is accomplished simply by the dividing of the parent into two; but among the higher plants and animals it is effected by the production of young, which by growth eventually resemble the parent. In the plant-world the **seeds** with which we are familiar are the young which are produced by the parents, and then cast off to begin life upon their own account.

The chain of life. Thus the plant-world displays, like the animal-world, an endless chain of generations, one succeeding another. In long-drawn procession the generations pass by. Did the children of each successive generation *precisely* resemble the parent, we should still see upon the earth forms long since passed away, and which we know only from their fossil remains. But plants and animals have an innate tendency to vary, and, as pointed out in the last chapter, any slight variation which is useful to the plant tends to be perpetuated. So the great march of the generations is accompanied by gradual change. The forests of to-day were preceded by ancient forests of a different aspect, and will in their turn be succeeded by fresh types. "The old order changeth, yielding place to the new." We shall see later on how important this question of successive generations is in relation to practical questions concerning weeds and their ways.

How plants feed and grow. As regards the manner in which plants live and build up their bodies, something has been said already, and very little more need be said here. In taking in food-materials, plants do not deal in solid matter; water is largely used, in which are dissolved certain necessary solids; the function of the roots is largely the collection of this water from the soil; roots are also useful for the anchoring of the plant in its place. But roots are not a necessary part of a plant's equipment; some of our common water-plants, for instance, have no roots, but float about freely, taking in water through the stems and leaves. The stems are the framework on which the leaves and flowers are borne, and spread out to the light and air; and they are the pipes through which the water collected by the roots is passed upwards to the leaves. The leaves are the main factories of plant food. They are covered with innumerable minute openings (stomata), by which materials can be taken in or given out. By means of the stomata several processes are carried on. The superfluous water collected by the roots, having been relieved of its valuable dissolved solids, is passed out into the air in the form of vapour. Carbonic acid, which is a compound of carbon and oxygen, is taken in, the carbon is used, and the oxygen is passed out again. A reverse process also goes on, for plants have to breathe just as animals do—that is, they absorb oxygen and give out carbonic acid. As regards nitrogen, it is an important constituent of the plant substance, but although the air is composed mainly of nitrogen, plants are not able to absorb it in its pure state, but take it in through the roots in the form of compounds with other substances—called nitrates. Thus the materials required to build up the plant body—carbon, oxygen, hydrogen, nitrogen and small quantities of various other substances dissolved in the water—are all assembled in the leaves. The green colouring matter of leaves (chlorophyll) has the strange power in the presence of light, of uniting these substances into the living material of which plants are made. The importance of the leaves is thus evident, and also the need that they should be widely spread out, so that all receive a share of light and air.

The making of the next generation. So far, the parts of a plant which we have considered—root, stem, leaf—are all concerned with providing for the life of the individual possessing them. In the flowers we come to that part of the plant which is set aside for the production of fresh individuals. The common type of flower consists of several rows of modified leaves arranged in rings. The outer ones (sepals and petals) are rather leaf-like, and some of them (usually the petals) are often brightly coloured. The inner and more important ones (the stamens and pistil) are often small and inconspicuous[1]. The pistil contains the germ of the future seed. In order that the seed may become mature, it is necessary that the fine dust called pollen, which is produced by the stamens, should be carried to the pistil. It is moreover generally desirable, and often necessary, that this pollen should be brought to the pistil from a different flower of the same species. This is often effected by the wind, which blows the pollen about from one flower to another; but in the more highly organized flowers insects are the agents used. These are attracted to the flowers by the honey which many flowers produce,

---

[1] The structure of the flower of the Buttercup is referred to on p. 64.

and also by the pollen, which is a useful food to them. The bright colours and delicious scent of flowers are also useful in attracting insects, and are characteristic of flowers which rely on insects, and not on the wind, to carry the pollen about. When the transfer of pollen has been safely effected, a tiny tube grows out from the pollen-grain and penetrates into the egg-cell, which is destined to form the seed. The showy parts of the flower fade, and by active growth within the ovary, or place where the young seeds are produced, ripe seed is formed. Then the seed-vessel opens or becomes detached, and the seeds are liberated from the parent plant. It is important to the species that the seeds should become well scattered. Underneath a large tree for instance, the shade is so dense that if its seeds fell straight down the young plants would probably be killed by want of light. And in the case of much smaller plants than trees, the parent may have used up much of the special materials in the soil which it needed for their growth ; so that it is desirable that the young plants should start life in fresh soil. We find in nature many ingenious and sometimes elaborate devices for the dispersal of the seeds of plants ; some of these not only ensure that the seeds are sown beyond the influence of the parent, but under favourable circumstances tend to transport the seeds over long distances. This question of seed-dispersal is of much importance in connection with the study of weeds, and a chapter will be devoted to its consideration.

When the seeds fall on favourable ground, and when they receive a suitable amount of heat and moisture, growth commences ; a tiny root protrudes from the seed-coat and turns downwards into the soil, and a tiny shoot emerges and turns upwards into the air. Both commence anew the gathering of food-material, and the seedling embarks in its turn on the life-cycle which has been already described.

Length of life. The period occupied by this cycle, from seed-stage to seed-stage, varies very much in different plants. While many of our common weeds run through their whole life in a few months, our noble

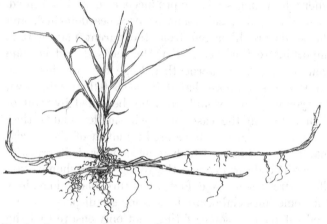

Fig. 2. Couch-grass (winter condition). Last season's upright shoot (with leaves); underground shoots arising from its base, terminating in buds ready to form upright shoots next season. ¼ nat. size.

forest-trees may live for hundreds of years. It is clear that this question is of much importance in dealing with weeds, for plants which seed rapidly and can therefore migrate quickly to and fro are not much interfered with, and are sometimes even encouraged, by man's digging and ploughing; other longer-lived and more sedentary species would soon be exterminated in the tilled land;

but sometimes they can settle down in the permanent pasture-lands and may prove very difficult to get rid of there, if they are deep-rooted.

Upright and prostrate stems. The majority of our perennial plants, whether they be herbs or trees, have stems which are upright. Such a stem may be

Fig. 3. Cuckoo-pint in early spring. Last season's creeping stem (on the left of the roots), now wrinkled and fading away. The coming season's creeping stem beginning to form to the right of the ring of roots. Reduced.

permanent, and very long, as in trees, or in the other extreme there may be an extremely short stem, from which annual stems spring, as in a large number of herbs, such as Monk's-hood, Agrimony or Onion. Often the stems, instead of being upright, are horizontal. Sometimes horizontal stems run along the surface of

the ground; in other cases they push their way like
roots under the surface. If as often happens they
keep sending out roots as they progress, there is no
limit to the length which they may attain. While the
length of an upright stem, such as that of a tree, is
limited by the question of sustaining its own weight
and the weight of the branches and leaves, and also by
its capacity to send up supplies to the leaves above,
there is no such limit to a horizontal stem which roots
as it runs, nor is there any limit to its life. Generally
such stems keep dying off behind as they advance, and
thus if they branch, each branch may become a separate
plant. This habit of growth is common among our
native plants. The Ivy and Strawberry are examples
of species whose stems creep above the ground; Couch-
grass and Bracken and Cuckoo-pint have underground
creeping stems (Figs. 2, 3). This habit of growth helps
some common plants to become very troublesome weeds,
and tends to make the plants grow in colonies. Thus we
may find Ivy filling a whole wood, and it is within the
range of possibility that the entire colony originated by
direct growth from a single seedling; and so with a
hill-side of Bracken. We shall have occasion to return
to this subject when we come to consider how weeds
spread.

EXERCISES II.

1. Collect specimens of any plants which you find able to grow
on stony ground such as a gravel playground, old walls, or quarry
rubbish-heaps. Name, draw or describe.

2. Examine any building sites which have been enclosed but
left unused for some years. What plants do you find there? Are
they annuals or perennials? Do the plants tell you in any way for
how long they have been growing?

3. Notice where Bracken grows. Is the Bracken-covered land flat or sloping, low-lying or lofty, clay or sand, wet or dry, shaded or sunny? Can you tell why Bracken grows there instead of grass or heather or gorse?

4. Read through a seedsman's catalogue of flower seeds. Copy out the names of any seeds offered of which the plants (e.g. Foxglove) already grow wild in your neighbourhood. Also note whether they are described as hardy annual, biennial or perennial.

5. Collect and dry the seeds of any interesting wild flowers. Put them up in packets neatly labelled.

6. Collect from a hedgebank several upright stems and several prostrate stems. Draw examples of each.

7. Dissolve some nitre (potassium nitrate) in water, not more than one ounce to a big garden watering can of water. Soak a small patch of grass lawn with as much of the solution as it can take up. Does it make any difference to the next fortnight's growth?

CHAPTER III

ON WEEDS IN GENERAL

Why weeds are harmful. We must now consider why the farmer and gardener wage such a constant war against weeds. The farmer's endeavour is to get as large a yield from each crop as he can. In a field of wheat, he tries to sow the seed so that each wheat-plant will have just enough room for its proper development, neither more nor less. If the seed is sown too thickly, then the plants will crowd each other, and all will suffer from insufficiency of soil or air or light. If the seed is sown too thinly, some of the available space will be unoccupied, and consequently will be lost. In a field of turnips similarly, he sows the seed sufficiently thickly in the drills to ensure a sufficiency of seedlings even in case of a partial failure of the crop. Subsequently he weeds out the superfluous seedlings, leaving just as many as the ground will most profitably support. Now, weeds not only occupy the ground which ought to be occupied by the crop, but they rob the crop of some of its food supply. The natural soil is generally not rich enough to nourish the heavy crops which the farmer desires, so at considerable expense he enriches it with manure of one sort or other—material which is especially rich in those mineral constituents, such as phosphates and nitrates, which plants most require. Weeds seize on this rich food with avidity, and steal it from the crop. Above ground a similar thing happens. The weeds crowd the young crop plants, and prevent their receiving their due amount of light and air. Weeds are also injurious because for the purposes of their own growth they rob the soil of part of its water supply. The effect of this is that, especially if the weather turns dry, the crop plants may be left short of water, and consequently their roots may not be able to send up into the leaves a full supply of the important materials which the plant receives dissolved in the water. The crop will then be unable to grow to its full size.

In these ways weeds can do a vast amount of harm if they are not rigorously kept in check, and their presence may result in a very serious money loss to the farmer. Prof. Pammel, writing of the United States of America, says, "A crop shortage on many farms in this country is in part due to the growth of weeds. Farmers everywhere could increase their crops by at least one-third by preventing the growth of weeds. The loss to farmers in every State would pay the taxes." He estimates that in the State of Iowa alone the loss of corn owing to the growth of weeds amounts to from seven million to nine million dollars every year. In Europe, an expert (Wollny) has placed the loss of crops in Bavaria due to weeds at the high figure of 30 per cent. In Mr Long's book on weeds the results of experiments with clean and weedy plots are given, in which the weedy plot in some cases yielded only one-half of the crop given by the clean plot.

It may be pointed out also that the pulling up and throwing away of large quantities of weeds does not make matters right, for many kinds of weeds absorb much of those substances in the soil which are specially valuable to the crop, and when we throw away the weeds we throw away these materials also. Once the weeds have grown, the damage is done ; what we need to do is to prevent the weeds from growing.

Weeds are also harmful because they harbour both animal and plant pests, which spread to the crops. The plant pests thus encouraged consist of minute fungi, known as rusts, which may do very serious damage to crops.

Then, when a crop of wheat or oats containing many weeds is cut and the grain threshed out, the seeds of the weeds will be mixed with it, and the farmer will get in consequence a smaller price for his grain, because it is impure. For whatever purpose grains are used—for making flour, for feeding animals, for distilling, or for sowing for next year's crop—they are much reduced in value by the presence of impurities in the form of the seeds of other plants.

Parasitic weeds. There is another group of weeds which may do very great damage to crops ; these injure them, not indirectly (like ordinary weeds), by robbing them of the raw food materials found in the soil, and of air and light, but by stealing from them the food which they have already manufactured. These *parasitic* weeds fasten upon the crop plants, and send into their stems, or into their roots, suckers which absorb the food material contained in the crop plants ; so while the parasites flourish and increase, the crop starves and withers. Such weeds do not do much damage in Britain ; they are less frequent there than in warmer and drier climates ; but two groups of them—the Dodders and the Broomrapes—are frequently found. Plants such as these, which do not manufacture their own food, have no need of leaves, nor of the green colouring matter (chlorophyll) by the aid of which plant-food is produced. The Dodders (*Cuscuta*) are small twining plants, yellow or red in colour, producing quantities of pretty little white waxy flowers. They are leafless and rootless, but send suckers into the stems of the plants they prey on, and although they are only annuals, they spread so fast that sometimes whole fields of clover are ruined by these voracious parasites. The Broomrapes on the other hand attach themselves to the roots of various plants, on which they form a knob, from which arises a leafless stem a foot or more in height bearing a number of two-lipped flowers, flower and stem being of one colour—brown, dull purple, and yellow predominating. The commonest species in Britain, the Lesser Broomrape (*Orobanche minor*), is of a purplish tint, fading to brown, and is frequent in clover fields especially in the south.

Poisonous weeds. The question of poisonous plants is also of importance in connection with our study of weeds. A number of plants contain substances of a more or less injurious nature. Among the flowering plants, species possessing such properties are found in many different groups. Sometimes one kind of plant is poisonous, while its near allies are quite harmless ; sometimes the hurtful quality runs through a whole group, as in the case of the Spurges and Aroids. The well-known Buttercup order possesses some highly poisonous species, such as the Monk's-hood (which yields the poison aconite), the Hellebores, and the common species of Buttercup, and almost all the members of this order are harmful. The order of the Umbellifers, to which the Carrot, Parsnip, Hemlock and Celery belong, also includes many highly poisonous plants. On the other hand, the great group of the *Compositae*, to which the Daisy, Dandelion and Thistles belong, include hardly a single hurtful plant ; and the Labiate order, which embraces so many highly aromatic plants, such as Mint, Thyme, Marjoram, Sage, Lavender, Calamint, and Balm, is also harmless. It is of interest to note that certain harmful plants are not equally poisonous to all kinds of animals ; for instance, slugs greedily devour various kinds of toad-stools of which a very small quantity would produce in ourselves violent sickness or even death.

If poisonous plants are allowed to increase in the farm-land, they may be eaten by cattle ; accidents of this kind occur frequently, and often have fatal results. When the seeds are poisonous, these may cause harm by getting mixed with the grain which is used for feeding animals, or which is ground up to make flour.

Beautiful weeds. It has been seen that weeds are simply plants growing where men do not wish them to grow. It follows that the common idea of weeds as ugly plants has no foundation ; beautiful plants may become weeds in our fields and gardens just as much as plants devoid of special beauty. Some of the most brilliant

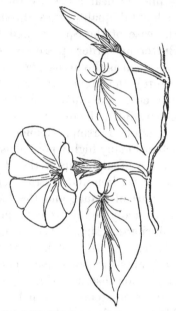

Fig. 4. Great Bindweed, one of the most beautiful of
native weeds. Reduced.

bits of colour that characterize our landscapes are due to weeds. Where in nature can we find anything to compare to the blaze of scarlet where, on light soils, Poppies are rampant among the corn? On similar ground, too, the Corn Blue-bottle (*Centaurea Cyanus*) sometimes provides patches of blue which in intensity can be

matched only with the Gentians of an alpine pasture.
A more characteristic piece of colour in Ireland is the
sheet of mustard-yellow where the Charlock has obtained
possession of the land; and in the western counties of
Ireland two weeds of pastures—the Water Ragwort and
the Purple Loosestrife—often combine to form in August
a glory of purple and gold to be equalled only by the
Heather and Dwarf Gorse of some of the stony mountain-
sides. Another common Irish weed which is always
welcome to the lover of beauty is the Corn Marigold,
or Gilgowan as it is called in the north, which during
the autumn and often until Christmas lights up our
potato fields with its yellow stars. Among graceful
plants what can excel the Great Bindweed? Some of
our prized garden-flowers are among the troublesome
weeds of other countries. For instance, a Californian
farmer would be amused to see the care with which we
grow the brilliant orange Eschscholtzia, which is to him
a familiar and unwelcome pest.

Weeds of different soils and situations. Every
child knows that different plants prefer different kinds
of soil and different sorts of places in which to grow.
We do not find Heather growing on clay, nor Ivy in
a bog; the plants of the sand-hills are not found in the
meadows, nor marsh-plants upon walls. If we take
wider areas, we find that the distribution of plants is
influenced likewise by climate. In woods in the mild
moist climate of Killarney, the little Filmy Ferns and
many rare mosses and liverworts cover the rocks and
tree-trunks; but they will not grow in such places in
the drier colder climate of Dublin. The weed flora is
effected in the same manner. About Dublin, and in the

maritime counties of Louth, Wicklow, and Wexford
there are lighter soils, less rain, and more sun, than in
most of the rest of Ireland. In these counties the weeds
that like light soils attain their greatest abundance.
Both climate and soil approach those found in the
English midlands, and the weeds of the counties
mentioned approach more nearly to those of England
than do the weeds of the west or north of Ireland.
These light-soil weeds, as was explained in Chapter I,
in many cases come originally from the Mediterranean
region, and their range has been artificially extended
by man northward and westward into France, or England,
or even Ireland. But in Ireland they are mostly found
on the light soils of the dry eastern counties, and they
decrease as we go westward. In 1866, while Poppies
were already abundant in eastern Ireland, no Poppy
was yet known to grow west of the Shannon. Since
then they have spread far to the westward—mostly, no
doubt, introduced by man along with agricultural seed
—but they are still very rare in the west, and their
increase as one goes by train in summer from Galway
or Sligo to Dublin will be readily noted by the observant
traveller. The Poppies are characteristic members of
this light-soil eastern group of weeds. Other weeds
with a similar distribution are the Penny Cress (*Thlaspi
arvense*), the White Campion (*Lychnis alba*), Welted
Thistle (*Carduus crispus*), Bugloss (*Lycopsis arvensis*),
and Red Hemp-Nettle (*Galeopsis Ladanum*). All of
these are well-known plants in fields and on road-sides
in the east, but they are unknown to the farmer in
Galway or Mayo. On the west coast of Ireland, on the
other hand, the prevalence of wet peaty soil and heavier
rainfall are reflected in the vegetation. In the meadows,

the Common Ragwort (*Senecio Jacobaea*) of the east
gives way to the moisture-loving Water Ragwort (*S.
aquaticus*), which is accompanied in the meadows and
pastures by a profusion of Purple Loosestrife (*Lythrum
Salicaria*). The Poppies, Valerianellas, Catchflies, and
other light-soil annuals are replaced by coarse Knot-
grasses (Polygonum). Yellow Flags (*Iris Pseudacorus*)
and even Royal Ferns (*Osmunda regalis*) become familiar
weeds in the poor pastures, and in the potato fields so
wet is the subsoil that the Reed (*Phragmites communis*)
which generally grows around the lakes in half a foot of
water, invades the ground, and sends up its leafy stems
among the crop.

Migration of weeds. It has been pointed out in
the first chapter that many of our corn-field weeds came
from the Mediterranean region, when the removal of
the forests and the ploughing of the land destroyed the
plant societies that hitherto had held the ground, and
allowed the fleeting annual plants to spread over the
country. Many of these Mediterranean plants themselves have probably come from the east, spreading
along the ancient trade-routes, helped by caravans
and by marching armies and their baggage-trains. A
European weed which we trample under foot may owe
its introduction there to Alexander or Hannibal, Caesar
or Alaric, and may indeed represent one of the few
remaining traces of empire-making marches and victories.
It is known that the Crusaders brought back with them
into Europe certain weeds hitherto unknown there.
Many of our commonest corn-field weeds, such as
Charlock, Wild Radish (*Raphanus Raphanistrum*),
Corn-cockle (*Lychnis Githago*) and Corn Blue-bottle

(*Centaurea Cyanus*) are foreign not only to our islands, but to Europe.

Three or four centuries ago, when roads were few and bad, and men as often as not marched across the open country on horse-back or on foot, the spread of weeds must have been slow. But recent engineering developments—notably the building of canals and of railways—offer safe and easy highways for water-weeds and plants of light soils respectively. In Ireland several light-soil weeds, which otherwise might never have spread beyond the coast towns, where they first landed, have taken advantage of the railways to spread along them all over the country. Two of these, the Slender-leaved Sandwort (*Arenaria tenuifolia*) and the Lesser Toadflax (*Linaria minor*), have long been known in England, though probably of Mediterranean or eastern origin ; a third, the Rayless Feverfew (*Matricaria discoidea*) is a North American plant, which appeared almost simultaneously in several parts of Europe late in the last century. The modern motor-car no doubt does its share in carrying weed-seeds on its

Fig. 5. Rayless Feverfew, an American annual. The flower - heads resemble those of a Daisy with all the white florets removed. ½ nat. size.

muddy wheels rapidly from place to place ; and now the aeroplane, alighting in a field and then flying a couple of hundred miles in a few hours, presents a possibility of rapid seed-dispersal, formerly possessed only by the migrating bird.

The increasing trade and intercourse between different parts of the world, which the advent of the steamship has brought about, have been accompanied by a copious interchange of weeds between different widely-separated countries. And while many of these invaders have been beaten back by the native plants, others have established themselves and spread rapidly. Thus, while the United States have sent several troublesome weeds to Europe, many of the commonest weeds of the north-eastern states are of European origin. In South America likewise, European plants are spreading rapidly. For instance, some British grasses (*Lolium perenne, Hordeum murinum, H. pratense*) are beating out the native grasses. Certain common British plants, introduced into Australia and New Zealand, are among the most persistent and troublesome weeds with which the colonial farmer has to contend. Among water-weeds, one of the most famous cases is that of the Canadian Water-Thyme (*Elodea canadensis*), introduced into Co. Down from America about 1836. It appeared in Great Britain a few years later and soon spread over the greater part of Europe, filling up rivers and ponds to such an extent as to interfere seriously with navigation and drainage. It has now decreased in abundance, and has ceased to be a serious pest, though still widespread.

EXERCISES III.

1. Collect and burn weeds until only the ash is left. What does the ash look like; has it any taste?

2. Take some of the ashes from a wood or turf fire. Try any experiments you can with them. Do they dissolve in water?

3. Chew and taste the leaves of Buttercup, Dandelion, Cowslip, Dead-nettle. Which are nice and which are nasty?

4. Which of the following weeds—Ragwort, Purple Loosestrife, Corn Marigold, Poppy—grow near your school? Can you account for their presence or absence?

5. Do you find any flowers, which do not grow elsewhere, particularly abundant along cart tracks, canal banks or railway sides?

6. Notice what plants remain uneaten round cow-byres or other places where cattle are often waiting.

7. Watch a goat. Write a list of all the different plants it is willing to eat. Will it eat the leaves of trees as well as herbs?

8. Make several maps of the fields within sight of the school. [This may be easily done by laying down the 25 inch Ordnance Survey on top of a dozen sheets of white paper. Prick through the corner points of the fields. Then rule in the field outlines in pencil.]

9. Watch the fields during the summer when they are sometimes golden with Buttercups, white with Ox-eye Daisies or pink with Persicaria. Watch how the colours change from week to week. Write on the map the colour of each field, the date, and the name of the flower to which the colour is due.

10. Paint the map to show the colours of the different fields on some particular day. Note the date—day, month and year.

11. Make drawings natural size of those flowers which give their colours to whole fields.

12. Paint these drawings.

NOTE.—Ordnance Survey Maps may be ordered through any bookseller. They are kept in stock by agents in the chief towns who can also show full catalogues.

The Dublin agents are Messrs Hodges, Figgis & Co., Ltd., 104, Grafton Street. The Edinburgh agents Messrs W. & A. K. Johnston, 2, St Andrew Square.

For particulars of sheets on 6 inch and 25 inch scales in England and Wales, see Index Maps, price 1*s*., from Edward Stanford, London, agent for the large scale maps of Ordnance Survey, 12 to 14, Long Acre, London, W.C. The 1 inch and smaller scale maps are obtained wholesale from Mr T. Fisher Unwin of 1, Adelphi Terrace, London, W.C.

For cheap edition for educational purposes only—price about 25 shillings for 200 copies—write to Director-General, Ordnance Survey, Southampton.

CHAPTER IV

SEEDS AND THEIR WAYS

In an earlier chapter we have seen how seeds are produced by the fertilization of the egg-cell contained in the flower. The production of seed is the crowning work of the plant's activity, for though by means of vegetative growth, such as is afforded for instance by creeping stems, some perennial plants can grow and spread indefinitely without the aid of seed, yet the production of seed is of the foremost importance in plant life. In the case of annual or biennial plants, seed-production is an immediate necessity, since in its absence a species might in a year or two become totally extinct.

The "seed" which the farmer sows, or tries to prevent being sown in the case of weeds, is sometimes the seed itself, sometimes the *fruit*—that is, the seed-vessel with one or more seeds enclosed ; or sometimes a portion of the seed-vessel. Thus, in the case of Charlock, Clover, Beans, the seed itself is sown ; but as we have seen or shall see, in the Buttercup, Bramble, Cow-Parsnip, Dandelion, &c., the fruit is sown. In the following pages, *seed* will be often used in its wide sense, signifying that which is sown, whether it be seed or seed-vessel. This is the sense in which it is used by the farmer and gardener.

Abundant Production of Seed. A very important point as regards the seed stage of plants is the fact that seeds are generally produced in great abundance. A creeping plant, which roots as it runs, may, in the course of years, produce, by vegetative growth, a hundred separate plants. But the same individual will probably, in a single season, produce a thousand or perhaps ten thousand or fifty thousand seeds, each capable of giving rise to a fresh plant. Let the student count the number of seed-vessels on the stem of a tall Foxglove, and count the number of small dark seeds contained in one of these seed-vessels, and by multiplying the two figures together, obtain the total number of seeds produced by this one plant, which was itself a single seed two years before. A plant of the Flixweed, an annual weed of sandy soils, has been estimated to produce 730,000 seeds. A single brown head of the Reed-mace or Cat's-tail (*Typha latifolia*) contains about a quarter of a million of seeds, each extremely light and furnished with a plume of delicate hairs, by means of which it may be wafted by the wind over a long distance.

Vitality of Seeds. The seed-stage differs from the

other stages in the life of the ordinary plant in several very important particulars. First, it is a period of arrested growth, and this cessation of vital effort may extend over a considerable period. Many seeds can certainly retain their vitality for years, and, although the stories of Mummy Wheat have no foundation in fact, still it is not impossible that some seeds may under favourable conditions be able to retain their vitality for very long periods of time. Next, this power of delayed growth is often accompanied by a remarkable power of resistance to injurious conditions. If we plunge an ordinary plant into boiling water for even a few seconds, it is absolutely killed; and severe frosts are likewise fatal to a great deal of plant life. But in the seed-stage plants are less sensitive. Many can resist a high temperature; on some seeds, the only effect of a quarter of an hour's boiling is to make them develop, when subsequently planted, faster than they would have done otherwise. In the same way, seeds have a great power of resistance to cold. Experiment has even shown that certain seeds have survived the almost unimaginably low temperature of liquid hydrogen ($-260°$ C.).

The advantage to plants of these exceptional powers of resistance in their seeds, and the bearing of these qualities on the question of weeds, are obvious. The chances of wide dispersal are increased greatly by the ample time-limit which is thus given. Little bits of the stems of many plants, if separated from the parent, will send out roots and grow readily if they are kept in contact with damp earth; in the case of succulent plants like the Stonecrops (*Sedum*) a bit of stem may be tossed about for weeks, and still grow as soon as it gets a chance; but seeds can endure the same treatment for months, or for years, and yet retain their power of growth. Again, a severe frost, or a great drought, or a fire, may destroy all the plants in an area; but some of the seeds which have been shed will probably survive, and re-establish the species when ordinary conditions return.

Again, were it not for the prolonged vitality of seeds, the extermination of annual weeds—and the majority of weeds are annual—would in most cases be a simple matter. If during one season we kept the ground thoroughly well weeded, so that none of the weeds produced seed, we might expect to see none of these weeds again, unless seeds or plants were introduced from outside. But the vitality of seeds renders this hope futile. With each spadeful of earth, with each stroke of the pick, we turn up buried seeds which have lain dormant perhaps for many years, but which are still ready to start growth under the stimulus of fresh air and light. It has been truly said that "One year's seeding is seven years' weeding." Indeed, the number of seeds lying in the topmost foot of soil all over the world, and especially in agricultural land, where ploughing and digging are continually burying fresh supplies, is truly enormous.

Abundance of Seeds in the soil. Some good examples of this are quoted by Long in his well-known book on weeds. "In a good garden soil which had been well cultivated for at least three years, few weeds having been allowed to shed their seed during that time, the author measured off 1 square yard and removed all the seedling weeds by hand on 17th May, 1909." The result may be summarized as follows:

	Number
Buttercup (? chiefly *Ranunculus repens*)	654
Annual Meadow Grass (*Poa annua*)	107
Dock (*Rumex* sp.)	60
Goosefoot (*Chenopodium album* ?)...............	26
Groundsel (*Senecio vulgaris*)	25
Others...	178
Total	1050

He also quotes some results of Korsmo, which are truly surprising. For instance, in a field intended for spring grain, the same crop having been sown for four successive years, no less than 33,574 weed seeds capable of germination were counted by successive weedings from one square metre (1⅕ square yard) of ground.

Let us take an Irish example. At Christmas, 1904, Mr John Adams removed the sod from about a square yard of a pasture field, undisturbed for 20 years, in Co. Antrim, and collected the soil which underlay the sod to a depth of 3 to 4 inches. He washed the soil, a large number of the seeds being lost in the process, and nevertheless obtained 829 seeds, belonging to 22 different kinds of plants. Eight of these species—Perennial Ryegrass, Sweet Vernal Grass, Curled Dock, Meadow Crowfoot, White Clover, Yellow Rattle, Ribwort, and Dandelion still grew in the field. Others were clearly the remains of cultivation, which had ceased over twenty years before—a seed of wheat, for instance, and abundant seeds of some common weeds, such as Orache (*Atriplex patula*), and White Goosefoot (*Chenopodium album*). Most of the other seeds belonged also to annual weeds of cultivated ground. Blackberry seeds were found, no doubt dropped by birds; also Birch seeds, probably carried by the wind. The seeds of several of the species showed themselves still capable of germination. Adams's observation shows that in the field which he tested there were in the top few inches of soil over four million seeds to the acre!

Thus we see that the soil must be looked on as a vast store-house of seeds, where are preserved the remains of many generations, and sometimes of a number of different kinds of vegetation. So long as the ground remains undisturbed, very few of these seeds have an opportunity of growing; but when the ground is dug or ploughed, myriads of seeds may spring into life. Thus the very operation of preparing the ground and sowing the crop as carried out by the farmer, often liberates the farmer's enemies in their thousands.

The Dispersal of Seeds. Merely by their small size as compared with the plants to which they belong, seeds are eminently suited to aid the spread of the parent species over the country. A seed is an extremely concentrated form of organism. The young plant is minute and tightly packed. Round about is generally some food material, also tightly packed; and the whole is enclosed in a tight tough coat. The seeds of most of our common plants are small and hard and roundish, and excellently fitted to survive many adventures. Seeds of this hard roundish type are borne by a large number of our common annual weeds—Charlock, for instance, and Chickweed, Shepherd's Purse, Poppies, Corn Spurrey, Hemp-nettle and so on. Such seeds, while they may possess great tenacity of life, have no special means of spreading themselves widely. Probably most of them were introduced originally in impure farm-seeds, and they continue from year to year (occasionally reinforced, no doubt, in the manner already specified), the parents shedding their seeds annually round about themselves.

But many seeds attain, owing to their own structure and to the action of external agents, a very wide dispersal, and we must now consider how this is accomplished. While certain seeds possess, owing to the expansion or contraction of appendages caused by dampness or dryness, limited powers of movement, none of the seeds of the higher plants can boast (as can the spores of some of the lower plants, such as sea-weeds) any power of locomotion. The seeds are themselves inert, and if they want to travel they must take advantage of some external medium or agent which moves. The most useful of all these agents is the restless wind. Water also is accountable for the transport of countless seeds. A third agent of the highest importance is furnished by moving animals, and particularly by birds.

Dispersal by Wind. The wind forms a very effective agent of seed dispersal inasmuch as all plants, except those which grow entirely under water or below the surface of the ground, are subject to its influence. The wind is responsible for the scattering to a greater or less extent of the seeds of almost all plants ; and though in many cases the scattering may not amount to more than a few feet, it is none the less of great importance as transferring the seed into fresh soil, and beyond the influence of the parent. The taller the plant, the greater will be the influence of the wind upon its seeds, both because the wind rapidly increases in force as one leaves the ground, and because it has a longer time to act on a falling seed; besides, a seed liberated above the main mass of vegetation has an increased chance of escaping entanglement among adjoining plants. With heavy-seeded plants borne on tall and strong stems, too, the wind helps to form a kind of catapult, the stems swaying to and fro and flinging the seeds out well clear of the parent. Acorns and Horse Chestnuts may be found often flung in this way clear of the shade of the parent trees; the same principle applies to many of the taller herbaceous plants whose seeds, when ripe, lie in cup-shaped seed-vessels.

When seeds are very small and light, catapult apparatus like the above becomes unnecessary, since the seed falls slowly enough to allow the wind to drift it far away before it reaches the ground. The seeds—or *spores*—of the lower plants, such as ferns, mosses, and fungi, are exceedingly minute and light, and are moreover borne in great abundance. For instance, a single Common Mushroom discharges into the air between one million and two million spores; and a plant of many of our familiar ferns produces each year spores far in excess of that number. These spores are so buoyant that they are carried everywhere, and are present in the air continually. We cannot leave a piece of cheese, or of bread, or an old shoe, exposed to the air—except under such dry conditions that there is no moisture available for vegetative growth—without their becoming blue-mouldy; this simply means that spores of the mould, which is a kind of fungus, which were floating in the air, have settled down and commenced to grow.

The seeds of none of the Flowering Plants are so minute or so abundant as these; nevertheless some attain a wide dispersal by the same means. The seeds

of Orchids and Broomrapes, for instance, are very small
and light, as are those of the Heaths and Rushes and of
the beautiful Grass of Parnassus. Such seeds may be
carried considerable distances by strong winds. In the
case of the Orchids, the seed proper, which is exceedingly
small, is surrounded by a loose netted envelope, which
tends to retard greatly the rate of fall. And this brings
us to the highly interesting group of seeds, which, while
sometimes comparatively large and heavy in themselves,
yet are capable of wide dissemination by wind on
account of certain appendages, which by increasing the
resistance of the air, cause the seeds to fall very slowly,
thus giving the wind a longer time to act upon them.
These appendages generally take the form of either
a membrane or wing encircling the seed or placed at
one side of it, or of a group of hairs, which also may be
arranged in one of several ways. It will be sufficient
for our purpose to take a few examples illustrating
leading types.

Wing Seeds. The Elm. In the Elm, the little
seed-vessel, which is flattish and
about $\frac{3}{16}$ inch in diameter, is surrounded by an oval membranous
wing about $\frac{5}{8}$ inch across, which
increases its area ten to twelve
times. The structure thus formed
falls slowly in an irregular zigzag
fashion, and is as easily carried
away by the wind as a piece of thin
paper of the same size might be.
Seeds of the same type are found
in the Yellow Rattle (*Rhinanthus*),

Fig. 6. Seed of Scotch
Fir (on left) and Elm
(on right). Natural
size.

and in some of the Umbellifers, such as the Cow-Parsnip (*Heracleum*) and Wild Angelica.

The Scotch Fir. Here the heavy egg-shaped seed, which is about ⅛ inch long, has a beautiful semi-transparent brown wing, shaped like the blade of a steamship's screw, or the sail of a wind-mill, extending for over half an inch on one side of it. The effect of this wing is that, when the seed is falling, a rapid spinning motion is set up, and the seed falls at only one-tenth or one-fifteenth of the rate at which it would fall were the wing not there. Thus, as a seed falls from the tree to the ground, the wind has ten or fifteen times as long to act upon it, and the chances are that it will be carried ten or fifteen times as far as it otherwise would. A familiar seed of the same type is that of the Sycamore.

Plume Seeds. Far more efficient, however, as well as more familiar, are those seeds which possess a plume of hairs or bristles. Such seeds are especially characteristic of the great order *Compositae*, which includes the Thistles, Dandelions, Groundsels, Sow-thistles, and so on, but they also occur in many other groups. In many of the most efficient of such seeds, the hairs are very long in proportion to the seed, and quite limp—as for instance in the Cotton-grass (*Eriophorum*), the Willows (*Salix*), Willow-herbs (*Epilobium*) and Sow-thistles (*Sonchus*). In many other plants the hairs are stiff, and are arranged in the form of a ball, or of a shuttlecock, or of an umbrella. Sometimes these hair-parachutes rise directly from the top of the seed, as in the common Thistles ; sometimes they are set on the top of a stalk, as in the familiar Dandelion (*Taraxacum*). Each of the hairs often bears numerous short branches, and a compound structure is thus produced which forms a very efficient air-drag. To take an illustration. If we were rowing a boat, it would give us very little trouble to tow a log of wood. But if for the log of wood we substitute a bush of the same weight, the resistance which the numerous branches and twigs offer to being dragged through the water would greatly diminish the speed of our boat. So it is with seeds which have a parachute of hairs, or *pappus*, attached to them. The well-known Spear Thistle (*Carduus lanceolatus*) has a comparatively heavy seed, to which a large and beautiful pappus is attached. The seed

Fig. 7. Seed of Dandelion. Enlarged.

alone falls quite fast, at least 15 feet in a second, but

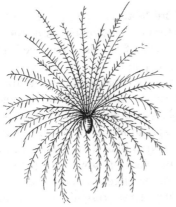

Fig. 8. Seed of Spear Thistle. Slightly enlarged.

with the pappus attached this speed is reduced to less than one foot per second. The pappus alone is so light that it falls less than half a foot per second; it is the pappuses of this and allied species, when detached from the seeds, that we see blowing about all over the country, and often entering railway carriages when we are travelling.

So far as the question of weeds is concerned, these flying seeds are of considerable importance, especially since they give to the plants which bear them a power of easy and rapid invasion from a distance. Our efforts to keep our gardens or fields free from such plants as Groundsel, Ragwort, or Thistles may be frustrated if a neighbour allows these to grow upon his land and to scatter their wind-borne seeds far and wide. Road-sides, too, which are often treated as a kind of no-man's-land, are a favourite haunt of many of these pappus-seeded plants, and unless they are kept clean the whole surrounding neighbourhood will suffer.

Dispersal by Water. The phenomena of dispersal of seeds by water need not detain us, as they only slightly affect the question of weeds. The seeds of about nine out of every ten plants sink in water, so unless in the case of water-plants, capable of rooting upon the bottom of ponds and rivers, water-dispersal is not of any very great use to plants. In low-lying districts, autumn floods may of course spread numbers of seeds over the ground, and no doubt marsh plants are widely dispersed by this means. The problem of how water-plants spread from pond to pond, or lake to lake, across intervening stretches of dry land, resembles in many ways the problem of the spread of land plants to islands. In some cases, as in the Reed-mace or Cat's-tail (*Typha*), water-plants use the wind to carry their seeds; but a more usual mode of dispersal is referred to in the next section.

Dispersal by Animals. As pointed out already, what plants need in order to attain a wide dispersal for their seeds is some moving medium by means of which the seeds may be carried far away from their parent. And while the wind, blowing where it listeth, and flowing water, play their part, roaming animals are also the conscious or more often the unconscious agents of much seed distribution. In flying animals such as birds, indeed, we have the most efficient of all means of dispersal, for while wind and water dispersal are limited by questions of the lightness or buoyancy of the seed, no such limit—excepting the case of very large seeds—arises in the case of birds. For instance, the heaviest of British seeds—the acorn—is sometimes eaten and sometimes scattered by Rooks intending to eat it.

There are two ways in which seeds may be spread by animals:—(1) they may become attached for a time to some part of the animal's body, or (2) they may be eaten, and subsequently dropped. To take the first of these cases:—small seeds may be embedded in the mud which adheres to the feet of animals, which thus unknowingly become dispersers of seeds. For instance, birds which, like the Black-headed Gull, often feed on ploughed land, may thus carry seeds. In a case within the writer's experience, twenty-one species of common field plants were found growing at a nesting-colony of this bird situated in the middle of a large bog in King's County. None of the plants grew naturally on the bog ; all had without doubt been brought by the birds from the adjoining farm-land. Again, the mud from the boots

of a man who landed on Clare Island in Co. Mayo after
a couple of days spent on the adjacent mainland, yielded
about fourteen different kinds of small seeds. More
interesting, however, are the cases in which the seeds
would seem to lay themselves out for dispersal by
animals. Many of our familiar plants bear seeds which
are provided with tiny hooks or prongs or barbs, by
means of which they may become attached, often very
firmly, to any passing animal. Thus, the seeds of the
Common Avens (*Geum urbanum*) and the Enchanter's

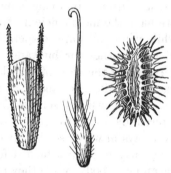

Fig. 9.   Seeds with hooks and barbs—Bur Marigold on left,
Avens in centre, Wild Carrot on right.   Enlarged.

Nightshade (*Circaea lutetiana*) have a long spine ending
in a hook. The seeds of the Bur Marigold (*Bidens*)
have each two barbed prongs, which make them quite
difficult to remove if they get into one's clothes; and
every child knows the round seeds of the Goose-grass
or Robin-run-the-hedge (*Galium Aparine*) all covered
with hooked hairs, and the fruit-heads of the Burdock
(*Arctium*) with their array of strong hooked bristles.
Seeds of this kind are especially fitted for dispersal by
wandering beasts, to whose hair or wool they become

attached ; the wool of sheep, for instance, is often full of seeds, and dogs may be seen sometimes biting themselves, where seeds entangled in their hair are annoying them.

Turning now to seeds which are dispersed by being eaten, we may limit our passing glance at this question to the case of birds, not because they are the only animals that disperse seeds in this manner, but because they are as a whole the principal eaters of seeds, and because, on account of their rapid flying movements, they are peculiarly efficient for scattering seeds widely, and for conveying them across stretches of water or other barriers which most creatures could not easily cross. Our native birds devour seeds of very many kinds in countless thousands, and while some, such as the Finches, Pigeons and Tits, grind up and destroy the seeds in the process of digestion, many others pass seeds through their bodies uninjured. Blackbirds and Thrushes, for instance, which feed extensively on berries, swallow the seeds (which are often enclosed in a hard coat), along with the juicy flesh, and the seeds are quite capable of germination afterwards. Without question, birds are responsible for a large amount of seed dispersal.

EXERCISES IV.

1. Write the date. Write in column a list of 20 common weeds. Say for each whether it is in bud, full flower, fading, ripening seed, ripe seed or whether seed has been shed.

2. Count the number of seeds in any of the following:—Gorse pod, Blackberry, Buttercup fruit, Dandelion head.

3. Even if you cannot count them, try to *estimate* the number of seeds on a whole plant of Foxglove, Willow-herb, Thistle or Ragwort.

4. Ill-treat some seeds [peas or mustard will do].   For instance pour boiling water over some, soak some in paraffin oil, mix some with poison (red lead), put some in the oven.   Next find out whether any of them are still able to sprout.

5. Find the time taken by different seeds to fall from a height, e.g. on a high staircase.   Take the time with a watch.

6. Examine the rubbish left by a flooded river.   Pick out any seeds or fruits.   How many sorts can you find ?   Draw each as seen through a hand lens.

7. What kinds of birds do you see eating garden fruits or wild berries?

8. Are any birds perching on the Thistle heads ?   If so, what birds ?   What are they doing ?

9. Collect some mud from a door-scraper or from cycle wheels. Can you grow any weeds from it ?

10. Examine your clothes after a woodland walk in autumn. Draw any fruits or seeds you have collected.

CHAPTER V

THE WAR AGAINST WEEDS

In the foregoing chapters we have learned a good deal about weeds and their ways, and we have now to consider how they are to be kept in check.   Let us summarize some of the facts bearing on their life-history.   First of all, an important distinction may be drawn between

(1)   Short-lived weeds (annuals and biennials),

(2)   Long-lived weeds (perennials).

Annuals. The short-lived weeds rely practically entirely on their seeds to carry on the race.   The

annuals root lightly, grow rapidly and seed rapidly and
abundantly, the whole life-cycle
often not exceeding a few months ;
hence they are specially fitted for
life in cultivated ground, which
is being frequently disturbed. If
we wish to get the better of
them, we have to be quick about
it. They mostly reach maturity
sooner than the crops among which
they grow ; hence the necessity of
weeding, or removing the useless
plants and leaving the useful ones;
and in weeding one of the main
difficulties often is that the weeds
must be left long enough to be
recognised as such, and to be con-
veniently handled, without being
left long enough to seed—and
often only a short interval sepa-
rates the two stages. Again, while
in many cases these annual weeds
ripen their seeds only during a
definite period—mostly the later
summer months—in other cases, in
our mild climate, they continue to
mature practically the whole year
round. Thus, if we wish to keep

Fig. 10. An annual plant
(Felwort). Note the
weak root-system,
short stem, and many
flowers — all charac-
teristic of annuals.
Slightly reduced.

our gardens clear of Groundsel (*Senecio vulgaris*) or
Annual Meadow-grass (*Poa annua*), there is not in
Ireland one month in the whole twelve during which at
least a little weeding must not be done. Another point
concerning annual weeds is this. They grow up very

fast, and as the commencement of their growth generally coincides with that of the crop (since the ground
is left alone after the crop is sown) they compete
directly with the young crop-plants for food and air
and light, and may succeed in smothering the crop if
we do not check them. They are generally easily removed, as their brief life does not require or allow of
the development of a deep or complicated root-system.

Biennials. Biennial weeds allow us a rather longer
respite. Many of them have formed by the end of their
first season a more or less conspicuous tuft or rosette of
leaves, and they pass the winter in this state, shooting up,
flowering, and seeding, during the succeeding summer.
Some of the most troublesome of the biennial weeds are
large plants—for instance Spear Thistle (*Carduus lanceolatus*) and Marsh Thistle (*Carduus palustris*). In the
winter state they are conspicuous, and can be destroyed
with a spud or by other means, or if cut down when
shooting up to flower in their second year they will die.
They often develop a long tap-root, as in the Wild
Carrot, and in any case, on account of their longer life,
have a better developed root-system than annuals ; so
they are usually more difficult to weed out than the
latter. Should biennial plants appear in the tilled land,
they are killed by the annual ploughing ; it is mostly in
permanent pasture that they effect a lodgment.

Perennials. With perennial weeds the case is
often different. Some, like the annuals and biennials,
are "spot-bound"; that is, they have no power of
creeping about and so spreading by vegetative growth.
The Ribwort Plantain (*Plantago lanceolata*) and Dandelion (*Taraxacum officinale*) are familiar examples.

These rely on their seeds for spreading. If we keep them from seeding by continual cutting, they cannot extend their range, but they often contrive to seed in spite of us. If we wish to exterminate them we must grub up the perennial roots ; and this must be done carefully, for many fleshy perennial roots have the power of sending out fresh shoots from any cut surface. Thus, if we cut a Dandelion root into a number of pieces, and plant them, each piece can grow and form a new plant.

Fig. 11. Dandelion (winter condition). ¼ nat. size.

Many of the most troublesome weeds of all, and the most difficult to deal with, are perennial plants with stems which creep about underground — Couch-grass (*Triticum repens*) for instance, and the Field Thistle (*Carduus arvensis*) and the Bracken (*Pteris Aquilina*). These plants rely only partially on seed for extending their domain. Their stems burrow rapidly through the soil, and even the softest of them, such as those of the Field Thistle or the Great Bindweed (*Convolvulus*

sepium) have a surprising power of penetration. These
creeping stems or rhizomes send up annually into
the air leafy stems, which not only manufacture the
annual supply of plant-food (see p. 15) but also produce
flower and subsequently seed. And therein lies the
weak point of this formidable class of weeds. It is
sometimes nearly impossible to exterminate their underground stems; these are often deep down, and every
scrap left in the soil has the power of growing and
founding a fresh colony; but by persistently attacking
their over-ground shoots, we may not only stop their
seeding, but by preventing them from manufacturing
food, we may starve them to death.

Colony-forming weeds. It is worth noting that
this creeping habit, by which a plant continuously
spreads from a centre, has a marked effect on its
general distribution as compared with those species
which spread by means of seed, and consequently advance by jumps. The former mode tends to form dense
colonies, the latter to produce scattered individual
plants. We see this well exemplified in the commoner
Thistles; the serried groves of the Field Thistle and
Meadow Thistle (*Carduus pratensis*) contrast strikingly
with the graceful isolated plants of the Spear Thistle
and Marsh Thistle. Similarly we may compare the
dense close tufts of the Common Rush (*Juncus effusus*)
with the extensive open colonies of the Sharp-flowered
and Blunt-flowered Rushes (*J. acutiflorus* and *J. obtusiflorus*). Other plants which form colonies on account
of their creeping stems will occur to the reader, such
as the Wood Anemone (*Anemone nemorosa*), Coltsfoot
(*Tussilago Farfara*), and many water-plants, such as

the Reed (*Phragmites communis*) and the Bullrush
(*Scirpus lacustris*). But it should be noted that there
are some familiar colony-forming plants whose increase
is due to seed, not to creeping stems—the Primrose
(*Primula vulgaris*) for instance, and the Ling (*Calluna
vulgaris*).

Bulbils. Another form of vegetative growth by
which weeds may increase—it is for-
tunately not a frequent method, for
it is rapid and very efficient—is by
the production, on the stem or leaves
or among the flowers, of buds which
become detached and at once grow
into new plants. Thus, the Crow
Garlic (*Allium vineale*) which is not
rare in sandy ground and on banks,
produces among its red flowers a
number of little onion-like buds,
which fall off and grow at once. The
common Lady's-smock (*Cardamine
pratensis*) often produces buds which
grow into young plants all over the surface of its leaves,
every one of which is ready to root and to begin life on
its own account.

Fig. 12. Flower-head
of Crow Garlic,
showing bulbils
among the flowers.
½ nat. size.

The dropping of the seed. Then as regards their
seeding, our war against weeds must be conducted with
a knowledge of the habits of the different plants in this
respect. Most of our annual weeds inhabit in their
native countries open ground, and they do not possess
in most cases any elaborate devices for scattering their
seeds widely. The seeds of Chickweed, the Speedwells,
Spurrey, Knot-grasses, and many others fall to the

ground and under ordinary circumstances are not
scattered more than a few feet from the parent. Other
seeds, such as those of the Cresses, Crane's-bills, Gorse
and Balsams, are shot out by the sudden rupture of the
seed-vessels, and may be thrown from 5 to 20 feet from
the parent. But the plants which produce seeds capable
of being carried by the wind have a great advantage
over the last group, and if not kept in check are a
serious menace to the farmer. One patch of dirty
ground, with its motley population of Thistles, Ragwort, Groundsel, Willow-herbs, and so on, can infect a
whole neighbourhood ; the flying seeds of these plants
go everywhere ; walls or fences will not avail to prevent
the ingress of these skilful aeronauts. Plants with
animal-borne seeds also deserve study, for here again
we find efficient long-distance dispersal in full swing.
Many of our bushes and trees, for instance, invade the
farm-land by the help of birds.

"Letting in the Jungle." In this connection we
must remember what we learned in the first chapter,
that forest is the natural condition of our country.
Were the people removed entirely from our islands,
there can be little doubt that a surprisingly short period
would see the native woods again established over much
of the farm-land, after a transition period during which
grasses and other perennial plants would hold the
ground. Even as it is, the native bushes and trees,
as well as numberless perennial herbs, are continually
trying to creep in. In tilled land, it is the periodical
ploughings that render these incursions futile ; in
pasture land, grazing has the same effect. But wherever
ploughs or spades, and cows or sheep, are excluded, we

Plate III

Glenariff, Co. Antrim

Looking down the glen towards the North Channel. The flat hill-tops (about 1000 feet) are occupied by peat-bog. The slopes and valley-bottom were filled with native woodland (Birch, Ash, Holly, Hazel, etc.), which has now been destroyed where the soil is rich to make way for tillage. On the steeper slopes the native trees are still dominant, and also along the banks of the stream; tillage and grazing alone prevent them from filling the glen once more. Introduced trees (Conifers) are seen in the left foreground. Photographed by R. Welch.

may watch the new birth of the native woodland. The Phoenix Park in Dublin is famous for its wealth of old Hawthorns, which no doubt are descended from ancient native stock ; but no young trees can be seen growing up to continue the picturesque Hawthorn groves ; the cattle and the deer see to that. But where new plantings of ornamental shrubs are made, and railed in to keep off the browsing animals, there young Hawthorns may be noted coming up in abundance. This is the work of the Blackbirds and Thrushes of the Park, which in winter feed on the haws, and afterwards, when sheltering among the shrubs, drop the seeds there.

Often in bushy ground, if we have eyes to see, we may study the whole process of the evolution of new woods. Birds, feeding voraciously on blackberries, drop their seeds among the grass. A young bramble, repelling by its thorny stems the grazing animals, succeeds in establishing itself. Its shoots arch over and, reaching the ground, their tips take root and form new plants, and a little grove of Brambles arises. Birds which have fed on Holly and Ivy berries, and the berries of the Mountain Ash, roost among the branches, and drop seeds ; the grove also acts as a net for capturing flying seeds, and in this way seeds of Ash and Birch are caught ; and these germinate and in time grow up, protected from the nibbling sheep by the thorny Bramble patch ; and so in time a clump of trees arises, which in course of time will combine with other clumps till a lofty wood may result at length. Read Kipling's story "Letting in the Jungle," in his *Second Jungle Book*, and you will realize in how surprisingly short a time, if man's fight against weeds be suspended, the native vegetation may sweep like a wave over the country, burying beneath its green billows every trace of his laborious husbandry.

How, then, is our war against weeds to be organized? This little book is not intended as a hand-book of agriculture, and therefore it would be out of place to discuss the relative merits of different kinds of tools or of the various chemical compounds—common salt, carbolic acid, sulphuric acid, sodium arsenate, copper sulphate, iron sulphate, and so on—which have been suggested or used as weed-destroyers. We can only indicate in a general way the principles which lie behind the farmer's practical acts. His activities as regards weeds may be divided into two categories: (1) the keeping in check of the weeds which are already present in the farm-land, and (2) the prevention of the incoming of new weeds.

Keeping the weeds down. As regards the first of these, a large amount of farm work is directed wholly or partly to this end. Ploughing and harrowing, in addition to their usefulness as regards the soil itself, are invaluable for the destruction of weeds of all kinds, perennial as well as annual. When ploughing is done before the weeds have ripened their seeds, it is much better than weeding, because, as was pointed out on p. 23, the weeds have taken up much of the nutritive materials from the soil. If we weed the land and throw the weeds away, as is too often done by careless farmers, all this material is lost. But if we plough the weeds down, they form a valuable "green manure." A similar result would be produced by burning the weeds and spreading the ash over the field, but in that case it is difficult to distribute it evenly over a large area.

Hoeing and raking are directed chiefly against annual weeds, which quickly die in the sun when their roots are exposed. When left lying, they rot and their useful constituents are re-absorbed by the soil.

Rotation of crops. Rotation of crops is useful as regards the keeping down of weeds because the different treatment required for different crops strikes at each kind of weed in turn. If oats are grown in a field year after year, for instance, the annual weeds will increase enormously, for they cannot be handled while the crop is growing, and by the time the crop is cut they have shed their seed. But if a root crop is taken off the land, an opportunity is afforded of decimating the crop of weeds. Similarly annual weeds will tend to disappear—temporarily at any rate—if the ground is laid down in grass; and the perennial weeds of pasture will be largely destroyed if the ground is ploughed and sown with an annual crop. So part of the farmer's policy is to keep chasing the weeds about, and not letting them settle down or get the upper hand anywhere.

Grass-land. In grass-land annual weeds get squeezed out by the stronger-growing plants, and the weeds are biennial or perennial. Two of the most familiar biennial pasture weeds are the Ragwort and the Marsh Thistle (*Carduus palustris*), which are referred to later on (pp. 81, 89). Such plants are best fought by cutting them down when they shoot up to flower in their second year. The perennial weeds include that especial pest the Creeping Thistle (*Carduus arvensis*); in damper ground such plants as Meadow Thistle (*Carduus pratensis*) and Meadow-sweet; and on the uplands the Bracken, and shrubs such as Gorse and Bramble. Most

of these will be also referred to later on. For such
perennial weeds as the Thistles, Meadow-sweet and
Bracken continual cutting is generally the best remedy
short of breaking up the land; weeds of the type of
Gorse and Bramble can be burned and the roots then
stubbed out.

Roadsides and Fences. But not only does the
actual farm-land need continual attention if the war
against weeds is to be successful; hedges, banks, roadsides and all waste places may become very serious
sources of infection if not kept clean. What is the good
of cleaning our fields of Ragwort if ten million or fifty
million seeds are allowed to mature on the adjoining
roadsides, every seed capable of being carried into our
fields by the wind?

If the land is to be kept clean our war must be
waged to the bitter end, and all persons interested in the
prosperity of the farming industry must combine in the
annual cutting of such weeds, to prevent their producing
"their savage wicked brood," as old Jethro Tull calls
them, that our farms may be kept clear of these pests.

Under a recent Act of Parliament (the Weeds and
Agricultural Seeds (Ireland) Act, 1909) it is now against
the law for an owner or occupier of land in Ireland to
allow certain of the worst weeds to grow; and he may be
fined heavily if after a warning from the Department of
Agriculture, Charlock, Ragwort, Coltsfoot, Thistles, or
Docks are found (of course in excess) upon his land.
If this law has the effect of getting all farmers and local
authorities to cut down regularly or otherwise destroy
the weeds upon their land, the prosperity of the whole
country will be increased.

Pure seed. Equally important with the keeping down of the weeds which are already in the farm-land is the prevention of the influx of new weeds through the agency of farm or garden seeds. In many cases no amount of care in saving and cleaning will produce absolutely pure seed ; but the amount of impurities can be kept down to a very low percentage. Until comparatively lately but little attention was given to this important matter. Seedsmen sold, and farmers bought, seed which was far from pure, and innumerable weeds were thus introduced into the farm-land. To repeat an instance from Long's work on Weeds, Dr Stebler has published a case where a sample of Clover seed, weighing 19·4 oz., contained 8478 extraneous seeds, representing 39 species of plants, most of which were weeds. In Ireland, until a short time ago, very impure seed was sometimes sold, and enormous quantities of weeds were thus introduced into the country. But the establishment in 1900 of a seed-testing station by the Department of Agriculture and Technical Instruction has greatly mended matters. All the leading seedsmen now sell seed of high and guaranteed quality, and for a nominal fee any farmer can have his seed examined and tested. The worst of all economies on the farm or in the garden is the buying of cheap and therefore probably bad seed. The chances are that either such seed is old, and has lost its germinating power, or else it is full of impurities, and will yield a swarm of noxious weeds.

EXERCISES V.

1. Weed the garden beds. Select specimens of deeply rooted weeds and of slightly rooted weeds. Shake or wash the roots clean of soil. Bring them indoors and draw the roots.

2. Find the longest Dandelion root you can. Measure its length in inches.

3. Mark out a square yard of grass on lawn or field. Pick out from it specimens of any plants which are not grasses. Name, describe or draw.

4. Mark out a hundred square yards (or less) of a pasture field. Count the number of Thistles on this area. Estimate the total number of Thistles per acre.

5. Describe the present condition of any piece of land once pasture which has been left ungrazed for a few years.

6. Compare the weeds in fields devoted to different crops—Oats, Potatoes, Turnips. Is any field more or less weedy than another? Is any sort of weed commoner with one crop than with another crop?

7. How can you distinguish between fields that have been used as pasture for years and other fields which are now pastured, but which have been used recently in other ways?

8. Find out what different kinds of weeds, both field weeds and garden weeds, a donkey is willing to eat.

9. Examine the close-cropped turf where sheep have been grazing. Can you make a list of the plants which they have been eating?

10. Describe the present condition of a neglected garden. Which weeds are most rampant? Which garden flowers are being overgrown or crowded out? Are any garden flowers running wild?

CHAPTER VI

SOME COMMON WEEDS

In the small compass of this book it will not be possible to treat of even the commonest weeds individually. All that we can do is to select a few—all of them common and well-known plants—and study them briefly. In this selection we shall endeavour to choose our weeds so that they shall represent some of the most familiar and most troublesome which are to be found in different kinds of ground—in corn-fields, in meadows, in gardens ; they will also be chosen so as to exhibit the different means by which weeds increase, and consequently the different problems which arise concerning their extermination ; and incidentally we may gain some knowledge as to their botanical characters, their affinities, and so on. But space will allow of only a very brief notice of each plant.

Nomenclature. A word may be said in explanation of the two Latin or Greek names by which, among botanists, plants are called. In old times a plant was specified by means of a Latin description which often occupied several lines. Linnaeus, towards the end of the eighteenth century, introduced the modern method of a two-word or *binomial* name. The first represents the *genus* or group of closely allied plants to which the plant in question belongs. Thus, all the Roses belong to one genus (*Rosa*), and all the Buttercups to one genus (*Ranunculus*). But all the grasses, though clearly allied to each other, are too diversified in character to be placed together in one genus : together they form a larger group, a *Natural Order* (*Gramineae*), under which the many kinds of grasses are arranged in a number of different genera. The second name distinguishes the particular kind of plant, or *species*. Thus, the Red Clover is *Trifolium pratense* (*pratense* means "belonging to meadows") and the White Clover is *Trifolium repens* (*repens* means "creeping"). The second or specific name of a large number of plants indicates as in these cases some characteristic feature. The first or generic name has been in many cases chosen for the same reason; in other cases it is the name by which the plant was called in ancient times; sometimes it is only a "fancy name." As regards the English names, many of these are Old English, and have come down to us from the French or German or Latin; but others are mere modern translations of the scientific names. The English names of many plants are misleading, and might deceive a beginner. For instance, Christmas Rose, Rock-rose, Guelder Rose, Rose of Sharon, Rosemary, are none of them roses, nor are any of them even closely related to roses.

CREEPING BUTTERCUP OR SITFAST, ᵱeᴀᵱbᴀ́n (*Ranunculus repens*).

The Sitfast is an extremely common weed in arable land and in gardens. It is easily known from other species of Buttercup by its habit of forming runners—long shoots which run over the ground, sending out here and there a tuft of leaves above and of roots below, each tuft forming in the second year a separate plant owing to the dying away of the runner, and sending out a series of runners in its turn. The result is a dense colony of strongly-rooted plants, spreading rapidly and capable of smothering out most other species. The Sitfast affords an excellent instance of the efficiency of vegetative reproduction. It also seeds freely, and so attains a wider dispersal than it would by runners alone. It is fond of wet soils, and is capable of growing where it is submerged during the winter months. On light soils it is less troublesome. It is a plant of much vitality, with unusually strong straight white roots, and difficult to weed out by hand. On the farm the harrow checks it effectually, while in the garden it must be forked or hoed out, or killed by trenching.

Two other species of Buttercup abound in grass-land. The Bulbous Buttercup (*R. bulbosus*) is fond of light

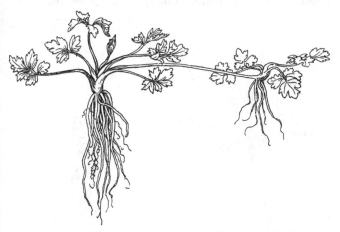

Fig. 13. Creeping Buttercup (winter condition). ¼ nat. size.

limy or sandy pastures. In appearance it resembles the Sitfast, but is easily distinguished by the bulb-like base of the stem and by the absence of runners. The Upright Buttercup (*R. acris*), which like the Sitfast favours damp land, is extremely common in pastures. It grows taller than either of the last, and is easily distinguished by its leaves, which, unlike the other two, are divided into quite narrow segments. All the species of *Ranunculus* (of which there are many others) are noxious, containing an acrid juice which is a frequent cause of cattle poisoning, and it is therefore worth taking trouble to keep down these plants on pasture land. The poison, however, is dissipated by drying, and in hay Buttercups form a nutritious food.

The Buttercups are useful to the beginner in botany as affording a simple study in the structure of flowers, since in them all the parts are separate, not joined together as in so many other plants. Note the outer-

Fig. 14.   Leaf of Upright Butter-   Fig. 15.   Base of stem of Bulbous
cup. ½ nat. size.   Buttercup. ½ nat. size.

most or lowest ring of the five green *sepals* (forming the *calyx*), which in the bud cover in the other parts of the flower and protect them. Next comes a ring of five shiny yellow *petals* (the *corolla*), each with a little honey-gland at its base. Both the bright colour and the honey are useful to attract insects, which in visiting the blossoms get dusted with the pollen, and fertilize the flowers (see p. 16) by carrying it from one to

another. Next comes a dense ring of *stamens*, the tips
of which (the *anthers*) contain the
grains of pollen; and finally filling
the middle of the flower is a group
of separate little *carpels*, each containing the germ of one seed. The
little beak at the top of each of
these is the *stigma*, the spot to
which the pollen must be brought
if it is to grow down and fertilize
the *ovule*, or young seed. When
the ovule is fertilized, it grows
into a seed which fills the hollow

Fig. 16. Ripe seed of
Creeping Buttercup. On
right a single "seed"
cut open, showing the
true seed within. En-
larged.

seed-vessel. The latter does not open to allow the seed
to escape, as in many plants, but drops off when ripe,
so what is sown is really a seed-vessel with a seed
enclosed.

The name *Ranunculus* is a diminutive of the Latin
Rana, a frog, from the wet places in which many of the
species grow; *repens* signifies "creeping." The English
name Sitfast is given from the well-known long tough
roots which render the plant so difficult to pull up. In
Irish the Sitfast is called ꝼeꝃꞃꞃán, derived probably
from an old word meaning a cow.

EXERCISES VI.

1. Collect specimens of all the different kinds of Buttercup you
can find. Each specimen should show root, stem, leaf and flower;
seed also if possible. Arrange them in bottles of water. Put to
each a label showing its name.

2. Draw a Buttercup flower in section many times natural size
as large as your paper will allow. Indicate the sepals, petals,
stamens, carpels.

3. Show by drawings how you distinguish the flowers of the different sorts of Buttercup. Explain in words the differences between the stems and leaves of different sorts.

4. Draw the seeds of different kinds of Buttercup much enlarged.

5. Press specimens of the different kinds of Buttercup in blotting paper or botanical drying paper.

6. Pick a nosegay of the most beautiful weeds you can find.

Botanical drying paper may be obtained from Messrs Watkins and Doncaster, 36, Strand, London, W.C. They also supply thick white botanical mounting paper in corresponding sizes.

COMMON POPPY (*Papaver Rhaeas*).

The Poppies are among the most decorative of "all the idle weeds that grow in our sustaining corn," and of the four species which so occur, *Papaver Rhaeas*, with its glorious full scarlet blossoms, is the most striking of all. The Poppies are characteristic members of the group of annual weeds that have followed cultivation from the warmer and sunnier parts of Europe; they love light soils, and are most abundant in the south-eastern parts of the British Isles, where sunshine and light soils are most prevalent. Thus, *P. Rhaeas* dies out as we go north into Scotland from England; and in Ireland, while it is abundant on the east coast, it rapidly becomes scarce as we approach the west. Four species of Poppy are found in our corn-fields, all somewhat similar as regards leaf and habit, but easily distinguished by their seed-vessels, even when these are quite young and are examined by pulling a flower-bud to pieces. Two of the four—*P. Rhaeas* and *P. hybridum*—have seed-vessels almost as broad as long; while the former is smooth, the latter is covered with stiff hairs. *P. hybridum* is besides a smaller and rarer plant, and has much less showy crimson flowers, each petal with a dark spot at the base. The two other species, *P. dubium* and *P. Argemone*, have *long* seed-vessels—about three times as long as broad. In *P. dubium* they are smooth; in *P. Argemone* covered with coarse hairs as in *P. hybridum*. *P. dubium* is the most widely spread of the Poppies, and is the Common Poppy of Ireland; *P. Argemone* is a rarer plant with smallish petals each with a large black spot at its base.

The Poppies produce a large quantity of small black

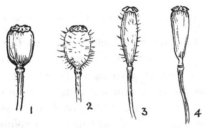

Fig. 17. Seed-heads of Poppies. 1, *P. Rhaeas*; 2, *P. hybridum*; 3, *P. Argemone*; 4, *P. dubium*. Slightly reduced.

seeds, which year by year give rise to a fresh generation. In light soils Poppies are often very troublesome weeds. Surface cultivation encourages them to germinate, and then they can be destroyed by the use of hoes and light harrows. Burying the seeds by ploughing does not help, as they retain their vitality and germinate when brought to the surface even after the lapse of years. All the Poppies are poisonous, and contain an acrid milky juice. The drugs morphine, opium and laudanum are prepared from this juice, obtained from the unripe seed-vessels.

If we compare a Poppy flower with that of the Buttercup which we examined recently, we shall find a striking difference as regards the seeds, which are borne in great numbers inside a single seed-vessel, instead of being arranged separately; and the calyx, formed of two large hairy sepals which drop off as the flower opens, is worthy of notice.

The name *Papaver* is supposed to be derived from *papa*, pap or thick milk, perhaps from the milky juice which the Poppies yield. The meaning of the name *Rhaeas* is obscure. The English *Poppy* is of course adapted from *Papaver*.

CHARLOCK, ꞃꞃᴀɩꞃeᴀᴄꞃ (*Sinapis arvensis*).

The well-known Charlock is perhaps the most universal and conspicuous of all weeds of cultivated land, and by competing with young crops for nutriment from the soil and for light and air, it is extremely injurious when abundant. Long, in his book on weeds, quotes cases where the yield of oats has been reduced by the presence of Charlock to $\frac{2}{3}$, and even to $\frac{1}{3}$, of the yield which was obtained when Charlock was not present. The hard round seeds possess great vitality, and can lie unharmed in the ground for half a century—possibly for much more—ready to germinate when an opportunity offers. Again, the young plants which spring up along with the corn ripen and shed their seed before the crop is cut, thus increasing the difficulty of keeping the weed in check. The cultivation attendant on root crops gives an opportunity for the wholesale destruction of Charlock; and for checking it among cereal crops, in addition to all possible hoeing, hand weeding, &c., spraying with

copper sulphate or iron sulphate has been found highly
beneficial. Like the Poppies, the Charlock is an annual,
infesting tilled ground and relying on its rapid and
abundant seeding for its continuance.

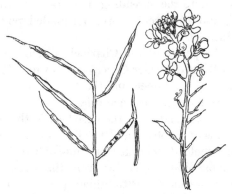

Fig. 18.   Flower and fruit of Charlock.   ½ nat. size.

The Charlock has two near relations which are also
frequent in cultivated land—the White Mustard (*Sinapis*

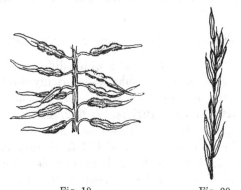

Fig. 19.   Fig. 20.
Fig. 19.   Fruit of White Mustard.   ½ nat. size.
Fig. 20.   Fruit of Black Mustard.   ½ nat. size.

alba) and the Black Mustard (*S. nigra*), the former in particular sometimes becoming, by its abundance, a very pernicious weed. The three are easily distinguished by their seed-pods, which differ as shown in the accompanying figure—note the spreading hairy pods of the White Mustard, and the erect short four-angled pods of the Black Mustard.

Note the petals of the Charlock, which are characteristic of the great order *Cruciferae* to which it belongs. They are four in number, and the two opposite pairs do not lie at right angles to each other.

A very near relative of the Charlock is the parent of all the forms of Cabbage and Cauliflower, and an allied species has yielded the different kinds of Turnip.

The name *Sinapis* comes from the Greek *Sinapi*, which was associated with allied plants ; *arvensis* = "belonging to fields." *Charlock* is *Cerlic*, the Anglo-Saxon name of the plant. The Irish name, ρ̇ρᴀιρєᴀċη, is derived from the Latin *Brassica*, Cabbage, to which genus the Charlock and its allies are closely related.

WHIN, FURZE OR GORSE, ᴀιτєᴀηη (*Ulex europaeus*).

Here we have a weed of a quite different character from those previously considered—a prickly shrub, which does damage by invading pasture land by means of seedlings, chiefly in hilly districts. It flourishes on poor dry or stony soils, which are sometimes not worth trying to reclaim from its dominion, and if left alone it can soon cover whole fields with an impenetrable thicket. In tilled land of course it has no chance, on account of the continual turning over of the soil ; but in pasture land, where other seedling shrubs would be exterminated by the nibbling sheep or cattle, the Whin effects a lodgment by means of its armament of spines, which keep animals from touching it. Otherwise it would not escape, as it is freely eaten by animals when the spines are destroyed by bruising. The plant is interesting in many ways. Note the apparent absence of leaves. Instead of ordinary leaves we find only a multitude of dark green spines, each ending in a sharp point. If we want to see normal leaves on the Whin, we must seek young seedlings. In these we shall find little leaves, each formed of three leaflets, like so many of the Clover family, to which the Whin belongs. But these leaves are not often seen after the first year, as they are rapidly superseded by spine-leaves, which fulfil the function of ordinary flat leaves in manufacturing plant-

Fig. 21. Branchlet of Whin. Slightly reduced.

food as explained on p. 15. The fragrant yellow blossom has the peculiar shape which we associate with the members of the Clover and Pea family—the *Leguminosae*.

Note the *irregularity* of the flower. Instead of having the parts, such as the petals, of the same shape and size, and disposed symmetrically round the stem, they are of different size and shape, and, as is common in irregular flowers, is set horizontally—not vertically, as in the Buttercup and Poppy. The petals of the flower are very irregular ; one large one occupies the top, forming a kind of roof ; two small ones are set one on

each side ; and two more are joined together by one
edge and form a kind of keel at the bottom. The
stamens and other parts are also strikingly irregular.

The seeds are borne in short downy pods. When
they are quite dry the pods split suddenly, throwing the
seeds violently out, and thus giving the seedlings a
better chance of life than they would have if they were
merely dropped into the middle of the thick bush.

Two other species of Whin (*U. Gallii* and *U. nanus*)
are found in the British Isles, the latter in the south of
England, the former more widely. Both are smaller

Fig. 22.   Seed-pod of Whin.
Nat. size.

Fig. 23.   Seed of Whin.
Enlarged.

plants than the Common Gorse, not so stiffly prickly,
and both flower in autumn.

Ulex is a name of obscure origin ; *europaeus* means
of course belonging to Europe. The English names
Gorse and *Furze* come to us through the Anglo-Saxon.
Whin is old English. In Irish the Whin is called
Aiceann, which may be related to the Welsh *eithin*,
meaning prickly.

BRAMBLE, Ɔꝶ, Ɔꝶeóᴣ (*Rubus fruticosus*).

Here we have another instance of a shrubby weed,
which, like the Gorse, succeeds in establishing itself in
pasture land in spite of grazing animals by being armed

with spines. In the case of the Bramble they consist of hooked prickles on the stems and on the back of the leaves, which are useful also in holding up the plant when it climbs through hedges and shrubs. It has two efficient means of increasing and spreading. The fruits, the well-known blackberries, are eaten largely by birds, and are thus carried about over the country ; the hard seeds which they contain are not injured by passing through the digestive organs, and when they are dropped

Fig. 24. Rooting tip of Bramble shoot (winter condition). Note the terminal bud turning up again, ready to start growth next season. Reduced.

Fig. 25. Seed (stone) of Bramble. Enlarged.

they are capable of growing. Again, the long shoots which are produced each year bend earthward in the autumn, and the tip takes root, and next year starts growth as an independent plant. In this way one plant may produce each year half a dozen fresh plants at a distance of 10 or 15 feet from itself. This tends to the production of dense groves of Brambles such as we see so often. As weeds, Brambles are formidable only on account of their armament of hooked prickles; they root lightly, and are easily stubbed out. A large number of different kinds, all very like each other, are to be found everywhere; but the Raspberry, the Dew-berry and the Stone Bramble are distinct, and easily recognized. All belong to the large order *Rosaceae*, in which are included almost all of our familiar fruits, such as Apple, Pear, Plum, Peach, Apricot, Strawberry, as well as many shrubs and herbs, such as Spiraea, Agrimony, Potentilla, Rose, and Hawthorn.

Rubus is the old Latin name of the plant; *fruticosus* signifies "shrubby" (some of the other species of *Rubus* are herbaceous). The English name *Bramble* comes from the old German name of the plant. In Irish Brambles and Wild Roses are both called ᴅᴙᴉᴙ (ᴅᴙᴉᴙᴇóᵹ in Co. Mayo), meaning perhaps "climber."

Mountain Willow-herb (*Epilobium montanum*).

The Willow-herbs are a group of native herbaceous plants, found mostly in wet places, ranging from sea-level to the tops of our highest mountains. Most of them keep to their own places, and except that some, such as the Great Willow-herb (*E. hirsutum*), tend to choke up streams and ditches, they are not mostly to be ranked as weeds. Two of them, however, the species at the head of this paragraph and its ally the Square-stalked Willow-herb (*E. obscurum*), are often found in tilled land, and are common garden weeds in many places. All the Willow-herbs send up erect mostly unbranched leafy annual stems, with four-petalled pink flowers which quickly change into long straight narrow seed-vessels, which split from the tip into four segments,

displaying within rows of small seeds, each furnished
with a delicate plume, by means of which it is wafted
by the wind and secures very wide dispersal. It is the
abundance of flying seeds that make these two Willow-
herbs a nuisance, since a few plants allowed to seed

Fig. 26. Mountain Willow-herb
(autumn state). The previous
season's stem is fading, and at
its base two buds have been
formed, which will produce the
stems of next year. Reduced.

Fig. 27. Ripe seed-vessels of
Mountain Willow-herb, one of
them opening and releasing the
seeds. Reduced.

can produce a whole crop of new plants over a considerable area. Otherwise they are easily kept in check,
since they increase slowly except by seed, and are
easily removed. The species we have chosen as an
example; *E. montanum,* is easily known from the others

by its smooth broad leaves. After the stem shoots up to flower in early summer two or three buds at its base give rise to little egg-shaped leafy shoots, which as the season goes on send out roots; the old stem dies after flowering, and the short shoots remain, to produce next year's stems. If weeded out early in the season, either these new shoots are not yet formed, or they are still firmly attached to the flowering stem, and weakly rooted, so that they come away with the old plant. But if we wait too long, when the old stem is pulled up they are well rooted, and break away and remain, and our weeding goes for nothing.

Fig. 28. Plumed seed of Mountain Willow-herb. Enlarged.

The name *Epilobium* is from the Greek *epi*, upon, and *lobos*, a pod, since the flower *appears* to be seated on the top of the ovary; the specific name *montanum*, "belonging to mountains," is not very appropriate. The English "Willow-herb" suggests the likeness of the leaves of some of the larger species (but not the flowers) to some of the Willows.

COW-PARSNIP, Uᴀᴘáɴ, ꝼuᴀᴘáɴ, ᴘléᴀʀɢáɴ, Sιuᴘáɴ
(*Heracleum Sphondylium*).

The Cow-Parsnip or Hogweed, with its large coarse hairy leaves and tall stems bearing flat umbels of white flowers, is a common perennial plant of meadows and moist woods. It is a characteristic member of the large order *Umbelliferae*, most of the members of which are easily recognized as such by the characteristic inflorescence, that is, arrangement of their flowers. From the top of the stem a number of short branches radiate (forming an *umbel*); each of these again divides at its

Fig. 29. One branch (umbellule) of the umbel of the Cow-Parsnip. Note the enlarged petals on the outer side of the outer flowers. Reduced.

summit into a number of branchlets, each of which bears a single flower. In both the primary and secondary set of branches, the outer ones are longer than the inner ones. The result is a large flat disk of close-set flowers (a *compound umbel*). In some plants the umbel is convex, like a ball; in some it is concave; in the majority more or less flat. The result of this grouping together of a large number of small flowers is to render them collectively very conspicuous, and consequently more easily found by useful insects which fertilize the flowers. In the case of the Cow-Parsnip this "advertising" of the flowers is helped by the fact that in the flowers round the margin of the umbel, where alone there is room for expansion, the outer petals are considerably enlarged. The seeds of our Cow-Parsnip are

Fig. 30. Single flattened "seed" of Cow-Parsnip. Enlarged.

also characteristic of the *Umbelliferae*, and deserve
study. The seed-vessel consists of a pair of carpels (in
this case much flattened, like little round plates)
adhering to a central axis. Each contains a single
seed: When ripe the carpels become detached except
at their upper extremity, and dangle for some time
before at last falling off.

Somewhat resembling the Cow-Parsnip, and also
common in damp meadows, is the Wild Angelica (*Angelica sylvestris*), but in this the leaves are composed
of separate finely toothed leaflets, the stem is smooth
and not coarsely channelled, the flowers are smaller, and
the whole plant smooth instead of being coarsely hairy.
Note in both these plants the beautiful tall column-
like hollow stems, combining strength and stiffness with
economy of material; these stems bear a considerable
weight of leaves and flowers, and, being taller than most
of the plants among which they grow, have to withstand
wind-pressure during storms. The Cow-Parsnip does
harm in meadows by shading and crowding out useful
plants such as grasses. It is not itself poisonous, though
belonging to an order which contains many highly
poisonous plants, such as Hemlock (*Conium macu-
latum*), Cowbane (*Cicuta virosa*), Water Dropwort
(*Oenanthe crocata*), Fool's Parsley (*Aethusa Cynapium*),
and Water Parsnip (*Sium latifolium* and *S. angusti-
folium*). The *Umbelliferae*, however, also include some
useful plants, such as Celery (which, however, contains
a poison except when bleached), Parsley and Carrot;
caraway seeds are the carpels of *Carum Carui*, and the
angelica of commerce is the candied stems of *Arch-
angelica officinalis*.

The name *Heracleum* is derived from Hercules, the

Greek god. *Sphondylium* is the name of an old genus
in which this plant and its allies were formerly placed.
The Irish name varies slightly—Sιuρán in Dublin, Uaρán
in Kerry, ꝼuaρán in Mayo. At Achill Sound, Mr Colgan
found pléaργán (i.e. ʒunna pléaργán, pop-gun, for which
its hollow stems are used by children).

CORN MARIGOLD (*Chrysanthemum segetum*).

Especially on light or peaty soils, the handsome Corn
Marigold (or Gilgowan, as it is called in the North of
Ireland) is often abundant—sometimes so much so as
to be a serious pest. It is an annual plant, possessed
of great vitality. If pulled up and thrown aside, it
continues to ripen and shed its numerous seeds ; and
in mild seasons the flowering period is prolonged, so that
one may sometimes gather it for house decoration until
Christmas. The plant is unmistakable, with its smooth
grey-green foliage and large yellow daisy-like flowers.
Note the leaves and stems, which have a waxy surface
on which water will not lie, but collects in little globules
like diamonds. Its flowers also demand our attention.
We have already seen, in the order *Umbelliferae*, as
represented by the Cow-Parsnip, how a large number
of small separate flowers may be borne close together,
and thus become conspicuous. In the order *Compositae*
this process is carried further. Just as most single
blossoms have a green calyx, which encloses and pro-
tects the flower while in bud, so in the plant before us
a number of modified leaves are collected to form a
calyx-like covering—called an *involucre*—for the group
of young flowers. The top of the stem is expanded into
a pincushion-like knob (the *receptacle*) on which a large

number of tiny flowers are tightly packed. These flowers consist, in the plant we are examining, of two kinds. Those of the cushion-like central part (the *disk*) are each perfect. The calyx is inconspicuous ; the corolla is tubular ending in fine spreading teeth, and enclosing the stamens and pistil. The outer row of flowers (the *rays*) are different. The tubular corolla is expanded on the outer side into a large petal-like *ligule* ; and the stamens are missing, the flowers possessing the female organ or pistil only. With certain modifications, this peculiar arrangement of the flowers is found all through the vast order *Compositae*, which numbers over ten thousand species, including all sorts of plants from trees to tiny herbs, and found in every country in the world. Sometimes the flowers are *all* tubular, as in the Tansy, Groundsel, and all the Thistle tribe ; sometimes they are *all* ligulate, as in the Dandelion and its numerous allies ; sometimes, as in the Corn Marigold, tubular in the centre and ligulate round the edge, either all of one colour, or the outer ones of a different colour, as in the Daisy, Sea Aster, and Chamomile.

The calyx, which in the plant we are examining is inconspicuous, in many of the *Compositae* attains a remarkable development, of which we shall have instances in several of the plants to be dealt with next.

The little seed-vessels are furrowed and blunt at both ends ; those resulting from the disk-flowers are roundish in section, while those of the ray are flattish

Fig. 31.   Seed (achene) of Corn   Marigold—from the ray.   Enlarged.

owing to the development of a corky wing on either side.

Chrysanthemum is derived from the Greek *chrusos*, golden, and *anthos*, flower, from the yellow flowers of some of the species; *segetum* signifies *(growing in) corn*. *Marigold* means St Mary's Gold, that is, St Mary's golden flower.

RAGWORT, buɑchɑllán buiohe *(Senecio Jacobaea).*

In the Ragwort or Ragweed we have one of the most abundant and pernicious of the weeds of pasture land. Growing vigorously during the first year of its life it forms a strong rosette of leaves which endures through the winter; and in the following season it generally shoots up, flowers, seeds, and dies. It is, like the Corn Marigold, a member of the great order *Compositae*, and if one of its flower-heads be examined, its affinities to the latter will be apparent. It differs in one character, which has an important bearing on the plant's

life-history. The upper or free portion of the calyx of each tiny flower is modified into a series of conspicuous slender hairs (the *pappus*), which remain attached to the seed-vessel after the corolla has faded and, acting as a parachute, cause the seed-vessel to be wafted by the wind far over the fields when it becomes

Fig. 32. Ragwort. Ray-flower (on left), disk-flower showing young pappus (on right) and seed. Enlarged.

detached from the parent plant. We have seen such a device already in the tiny seeds of the Willow-herb (pp. 75, 76), but it is in the *Compositae* that the pappus attains its most frequent and most elaborate

P. 6

development. In the Dandelion and Spear Thistle, to be dealt with shortly, we shall study two cases of a beautifully developed pappus.

The handsome divided leaves of the Ragwort need no description. The flowering stem is tall and tough, and owing to the arrangement of its numerous branches, all the flower-heads are borne at about the same level, producing, as in the umbel of the Cow-Parsnip (p. 77),

Fig. 33. Leaf of Ragwort. Reduced.

a far more conspicuous display than they would otherwise do.

Since the Ragwort spreads, and spreads very widely and abundantly, by means of its flying seeds, the way to check its increase is to prevent its seeding. This is best done by cutting it down where it shoots up to flower and so becomes conspicuous and open to assault. But we should note that the plant possesses great vitality, and if cut down in flower, or should damp

weather prevail, it may ripen its seeds in spite of us.
Another useful means of extermination
is furnished by the fact that, although
cows will not touch it, sheep devour the
plant with avidity, especially in its succulent earlier stages, and will soon clear
a field of Ragwort if turned into it.
There are nearly a dozen members of
the genus *Senecio* native in the British
Isles, one of the best-known being the
Groundsel (*S. vulgaris*). The one which
most resembles the Ragwort is the
Water Ragwort (*S. aquaticus*). It
grows in wetter places, and has much
less divided leaves and fewer larger
flower-heads.

Fig. 34. Leaf of
Water Ragwort.
Reduced.

Senecio is derived from the Latin
senex, old, from the abundant hoary
pappus; *Jacobaea* is from Saint James
(*Jacobus*), perhaps because he is the patron saint of
horses (as a cure for diseases of which the Ragwort
was formerly employed), or because it flowers near the
Saint's day. The Irish name is ᵬuᴀchᴀᴌᴌán ᵬuιᴏhe.
ᵬuᴀchᴀᴌᴌ, now "boy," formerly meant "herdsman," so
the name signifies "little yellow cowboy," probably from
its being so conspicuous in pasture land.

DANDELION, Seᴀᵽᵬhán, Cᴀιᵽ-cᵳeᴀᵽᵬhán (*Taraxacum
officinale*).

The familiar Dandelion is a native plant which
invades every corner of man's domain—tillage, pasture, garden plots, paths, roadsides, and is difficult of

extermination on account of its long tap-root (see Fig. 11) and its numerous flying seeds (see Fig. 7). The succulent root sometimes attains a length of several feet, going straight downward ; and if cut in pieces, each portion is capable of sending out shoots and giving rise to a fresh plant. Note the flower-heads. Their characters are those of the order *Compositae*, but unlike those of the Ragwort and Corn Marigold they do not possess a central disk of tubular flowers, but all the flowers are ligulate, the corollas having a long finger-like limb which is directed outward from the centre of the flower-head. Note also the beautiful parachute arrangement which crowns each little seed-vessel when ripe—a delicate stalk ending in a radiating group of hairs, acting most efficiently as a seed-distributor (see Fig. 7). While the seed is ripening the head is sensitive to moisture, so that in damp weather the involucre closes tight round the seeds, keeping them dry, to open again in sunshine, when at length the seeds are launched forth on their balloon journey.

To keep Dandelions in check we must cut them off deep down, or the roots will only sprout afresh ; and at all cost prevent seeding by knocking off every flower-head. Spraying with solutions of iron sulphate and copper sulphate has been tried with some success in Germany and in America for dealing with them on a large scale.

The Dandelion has got important medicinal qualities, and when bleached the leaves form a pleasant and wholesome salad.

The name *Taraxacum* is an old name of Arabic or Persian origin ; *officinale* means "medicinal," from the well-known qualities of the plant. Dandelion is a corruption of *dent de lion*, lion's tooth, from the tooth-like segments of the leaf. In Irish the Dandelion is known as Se아rbhán, little bitter [plant], or Caiꞃ-tꞃeárbhán, curled little bitter plant. The name is of course derived from the bitter juice.

Fig. 35. Coltsfoot. Note the underground stem, rising to the surface to send up flowers. Reduced.

COLTSFOOT, Aóhann, Sponc (*Tussilago Farfara*).

The Coltsfoot and its allies the Butterburs (*Petasites*) form a little group of the *Compositae* which are very persistent weeds. Their stems creep about underground, sending up so dense a growth of broad leaves that hardly any plant can live in their company. Their creeping habit tends to the production of large patches, on which they reign supreme. The Coltsfoot is fond of disturbed ground—railway banks, roadsides, and tillage. Its flowers appear in earliest spring, before the leaves—yellow heads with many narrow ray-flowers enclosing a compact disk, borne on leafless scaly stems. The leaves which follow are green above, white and felty below, 6—12 inches high. The seeds are distributed widely on account of the well-developed pappus. The creeping underground stems grow fast, and are white and numerous.

Its ally the Common Butterbur (*Petasites vulgaris*) is a much larger plant, with thick fleshy creeping stems. As in the Coltsfoot, the flowers appear before the leaves. They are pinkish, borne on short fleshy stems in a conical or cylindrical mass, in early spring. The rhubarb-like leaves are sometimes in wet ground as much as six feet high and three feet broad, forming an impenetrable shade under which nothing can grow. A smaller species of Butterbur, the so-called Winter Heliotrope (*P. fragrans*), a native of the Mediterranean region, is now one of the most troublesome weeds in many gardens and shrubberies. In size and growth it resembles the Coltsfoot, but its leaves, instead of being angular and white below as in that plant, have an unbroken margin and are green underneath. Its heads of fragrant purplish-grey flowers are borne in the later part of the winter, and are closely followed by the leaves. Its stems creep rapidly underground, and are most difficult to eradicate.

Plants such as the Coltsfoot and the Butterburs

are best kept in check by continual cutting. A cutting when in flower prevents their spreading by seed, and the mowing down of the young leaves prevents sustenance being conveyed to the underground stems, and weakens or kills the plant.

The name *Tussilago* comes from the Latin *tussis*, a cough, from its use as a cough medicine. *Farfara* recalls the resemblance of its leaves to those of the White Poplar, an old name of which is *Farfarus*. *Coltsfoot* refers to the shape of the leaves. The meaning of the Irish names of the plant, Ɑohann and Sponc, is obscure.

SPEAR THISTLE, ϝeóchɑoán (*Carduus lanceolatus*).

The Thistles are among the handsomest plants in our flora, and of them none is handsomer than the Spear Thistle, with its bold spiny foliage and large heads of pink-purple flowers. It is a biennial plant. By the end of its first season's growth it has formed a rosette of formidable spiny leaves, which remain through the winter ; next season it shoots up, producing a tall branching stem with numerous flower-heads. The flowers are succeeded by little

Fig. 36. Flower-head of Spear Thistle. ½ nat. size.

seed-vessels surmounted by a large and beautiful pappus (see Fig. 8), by whose aid they are widely dispersed by the wind. After seeding the plant dies. The farmer's

business is to keep the plant from seeding, and this is
done most easily by cutting it down when it shoots up

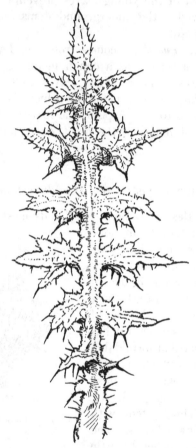

Fig. 37. Leaf of Spear Thistle. ½ nat. size.

to flower. The Spear Thistle is magnificently armed
against intruders. The leaves are beset with most

formidable spines; the stem is furnished with wings which are likewise spiny: the flower-head is amply protected, each of the leaves which make up the involucre ending in a spine. We are not surprised to see the Spear Thistle standing secure and untouched where all the surrounding herbage is closely nibbled by sheep and cows. Equally common in pastures, but preferring wetter ground, is the Marsh Thistle (*C. palustris*), a tall plant with a less branched stem, leaves smaller and darker, flower-heads much smaller, the whole plant armed with small spines. It is also biennial. The worst pest in all the Thistle group is the Creeping Thistle (*C. arvensis*) which has far-spreading brittle fleshy underground perennial stems. This habit of growth causes it to form dense colonies of annual leafy stems ending in smallish pink-purple flowers. Perfect seed is produced but seldom, but in spite of this the plant is only too common. The leaves are prickly along the margins.

On account of its perennial nature and creeping habit, the Creeping Thistle is extremely difficult to eradicate, and is one of the worst of all weeds. The best plan is to starve it out by continual cutting. Steady cutting throughout two seasons is stated to free the land of this pest. The presence of Thistles in any crop, hay or cereal, renders harvesting difficult and unpleasant; these inconveniences have to be added to the harm which Thistles do in robbing the soil and shading and competing with useful plants.

The meaning of the name *Carduus* is doubtful; *lanceolatus* means narrowly elliptic and pointed at both ends, referring to the shape of the leaves. *Thistle* is the Old German name for these plants. In Irish Thistles are called ꝼeóċaᴅán, "probably from ꝼeóᴅh, 'withering,' from the conspicuous appearance of withered thistles in a grazed pasture " (Prof. MacNeill).

GREAT BINDWEED (*Convolvulus sepium*).

This is a very handsome plant and a very pernicious weed. It produces underground a network of creeping stems, white, brittle, and succulent, every broken piece of which is capable of growing. From these arise the long leafy twining stems we know so well, with their beautiful white trumpet-shaped blossoms (see Fig. 4).

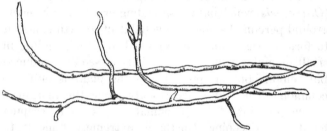

Fig. 38.   Underground stems of Great Bindweed (winter state).   Note the buds ready to send up next season's shoots.   ½ nat. size.

It is in the garden that this plant gives most trouble. Once its underground stems get established in places which cannot be dug, as among the roots of Gooseberries or of perennial clumps in the herbaceous border, it cannot be removed ; each year its crop of twining stems overwhelms the plants, and underground it pushes boldly out in all directions. Continual cutting and weeding alone will keep it in check, unless where we can clean the ground thoroughly of its white twisted stems.

A smaller species, the Field Bindweed (*C. arvensis*) is sometimes an injurious weed on light soils. Its leafy

stems are seldom more than a couple of feet in its
length, and its abundant trumpet-shaped flowers are
white or pink, often displaying both colours. Its habit
of growth is the same as that of the Great Bindweed.
Root crops accompanied by persistent hoeing are recommended as the best means of reducing it. A third
kind of Convolvulus (*C. Soldanella*) is found on sandy
shores in our islands, and is one of our most beautiful
wild-flowers. Its leaves are small and kidney-shaped,
and its great pink and white blossoms barely rise above
the sand on which the short stems lie.

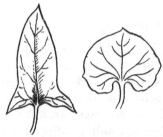

Fig. 39. Leaves of Field Bindweed (on left), and of Sea Bindweed
(on right).

The mechanism by which such plants as the Bindweed twine is extremely interesting. While plants in
general possess certain limited power of movement,
climbing plants display movement to a very marked
degree. In the Convolvulus, the upper part of each
growing stem is continually moving slowly round in a
circle. Should a stem in its rotation meet with say a
branch or twig, it is arrested at the point of contact ;
the portion of stem above the point of contact continues
to rotate, with the result that the stem becomes wrapped
round the branch, and as the Convolvulus grows the

wrapping round continues as long as growth goes on and no fresh obstacle intervenes. The advantage to the Convolvulus is obvious, as by this means its leaves and flowers are lifted high into the light and air on the shoulders of the plant to which it has clung; but the latter may suffer severely by being smothered under the climber.

The plant gets its name from the Latin verb *convolvere*, to entwine; *sepium* means "of hedges," from the habitat in which it is so often found. The English *Bindweed* also refers to the plant's characteristic habit of growth.

Marsh Woundwort (*Stachys palustris*).

In Ireland this is often an extremely common weed of arable land, but is not so much so in England. It is a native of wet places, whence it migrates into the tilled land especially in boggy districts. It is easily recognized by its square stem, strap-shaped hairy leaves in opposite pairs, and dull two-lipped irregular purple flowers disposed in ring-like form above each pair of leaves on the upper part of the stem. The plant runs about underground, sending up these annual leafy flowering stems. The "seeds," which are black and shining, and of which when ripe four may be seen grouped at the bottom of the calyx-tube, are really little nut-like fruits containing each one seed. Another species of Woundwort, *S. sylvatica*, which occasionally gives trouble as a weed, has its home in woods and copses. In flower and habit of growth it resembles *S. palustris*, but is easily known by its leaves, which are stalked, heart-shaped, and broad.

The Woundworts are characteristic members of the

large order *Labiatae*. Almost all the *Labiatae* are strongly scented plants. In the Woundworts the scent is rank and unpleasant, but their allies furnish to our

Fig. 40. Marsh Woundwort. ½ nat. size.

Fig. 41. Leaf of Hedge Woundwort (*S. sylvatica*). ½ nat. size.

gardens a large number of fragrant herbs—Mint, Thyme, Marjoram, Lavender, Sage, Calamint, and Balm.

The plant takes its name from a Greek word signifying *a spike*, from the way in which the flowers are borne; *palustris* signifies pertaining to marshy places. The English name refers to the supposed healing powers of the plant, which had a high reputation in old days for the curing of wounds.

Persicaria or Redshank, Glúineach dearg (*Polygonum Persicaria*).

Over a dozen Polygonums are native in the British Isles, but the above species is the only one which is ever a serious pest. It is a coarse-growing branched annual plant, well distinguished by its stems with red swollen joints, its leaves with a dark patch in the centre, and its cylindrical clusters of pink flowers. It is a late flowerer, sometimes making the autumn fields quite bright with its small blossoms. The Polygonums are closely related to the well-known Docks, and these two genera, along with the Mountain Sorrel (*Oxyria*), constitute the British representatives of the order *Polygonaceae.* The flowers of our plant differ in some respects from those of any of the other weeds we have been considering. The calyx and corolla are not distinguishable; they are represented by a five-parted corolla-like structure. This does not fade, and eventually clasps the little seed-vessel, which is three-angled, dark, and shining, and contains a single seed. To keep Persicaria in check, the main point, as in the case of all annual weeds, is to prevent seeding; this is accomplished by energy in hoeing and hand pulling.

Its relations the Docks are very troublesome weeds, the most abundant being the familiar Broad-leaved Dock (*Rumex obtusifolius*), with large broad flat leaves, and the Curled Dock (*R. crispus*) with long narrow wavy leaves. These plants have very long strong tap-

Fig. 42. Persicaria. ½ nat. size.

roots, any portion of which will produce buds and start growth; and they are also very abundant seeders. Autumn hoeing is very useful in destroying seedling plants; the old plants must be removed by pulling up

when the ground is soft (luckily the tall flowering stems
are tough) ; if the root be cut, this must be done very
low down, or it will sprout. The old plants when removed should be burned, as if thrown aside they will
probably go on growing.

The scientific name is from the Greek, signifying
"many joints," and refers to the knotted stems, whence
also the English name of the genus, *Knotweed* ; *Persicaria* is from the Latin *Persicarius*, a peach-tree ; the
plant is sometimes called "Peach-plant." "Redshank"
refers to the red knotted stems. The same red knots
on the stem give to the plant its Irish name ᵹlúıneᴀċ
ꝺeᴀƞᵹ, the first word referring to the "knees," and the
second to their colour.

NETTLE, neᴀnꞇóᵹ, lᴀnnꞇóᵹ (*Urtica dioica*).

The Nettle is one of those plants which are followers
of man. Although it is never deliberately planted, and
although its seeds are but rarely found in seed which
is brought into the country, it is widespread wherever
cultivation extends, and in the most remote spots where
it occurs it is usually found in connection with some
ruined hovel or with signs of former cultivation. Each
season each flowering shoot of the Nettle sends out
several prostrate stems, which creep on or just under
the surface of the ground, and give rise to the flowering
shoots of the following year. The result is a tangle of
tough matted stems, which, as they are strongly rooted,
are not easily removed. Excepting the Roman Nettle
(*Urtica pilulifera*), a rare plant of eastern England, only
one other species grows in the British Isles. This is
the Small Nettle (*U. urens*), an annual plant usually

found on roadsides and near houses ; it is easily known from the Great Nettle by its shining rounder leaves, short flower-clusters, and annual character. The small green flowers of the Nettles are none of them perfect ; some are male (bearing stamens), some female (bearing a pistil).

The Nettles are unique among our native plants as possessing stinging hairs, the efficiency of which is only too well known to all of us. These hairs consist each of a single cell, delicate and tapering, springing from a

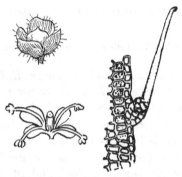

Fig. 43. Common Nettle. On left female flower (above) and male flower (below), enlarged. On right stinging hair, much enlarged.

little cushion of cells belonging to the epidermis or skin of the leaf. Each is filled with a poisonous fluid in which formic acid—the poison which makes the bites of ants so painful—is a leading constituent. Each hair ends in a little knob, which is bent to one side ; and below this knob the walls of the hair are very thin. As a consequence of this, when anything presses against the hair the knob breaks off, and the jagged end of the hair, filled with poison, penetrates the body which has pressed against it, if it is at all soft, injecting the poison through the hollow tip.

Continual cutting will soon weaken groves of the Great Nettle, but it is better, where possible, to root them out altogether; though tough and well-rooted, the stems and roots are wholly close to the surface.

The name *Urtica* is derived from the Latin *uro*, I burn, from the effect of the stings; *dioica* recalls the fact that the male and female flowers are borne on separate plants. *Nettle* is from the Old German name of the plant. In Irish the plant is called neantóʒ (in Kerry Lanntóʒ), a word of obscure origin.

Sun Spurge (*Euphorbia Helioscopia*).

Two kinds of Spurge, the Sun Spurge (*E. Helioscopia*) and the Petty Spurge (*E. Peplus*) are common weeds in arable land and gardens, and a third, the Dwarf Spurge (*E. exigua*) is sometimes also abundant, chiefly on light soils. All three belong to that class of annual weeds which claim the farmer's hospitality and are seldom found except in fields and gardens, and all of them have probably an eastern origin, ranging as they do throughout the Mediterranean region and on into India. A number of other species of Spurge are found as natives in the British Isles; they are mostly perennial plants, growing in woods or on the sea-shore.

The Sun Spurge is a striking plant. Its leafy stem, 6 to 12 inches high, terminates in an umbel-like group formed of leaves (of a pretty golden hue) and flowers, borne on five radiating branches which are themselves branched. The flowers are peculiar and interesting, and their structure may easily deceive a beginner. What

we might take for a single small flower is really a group
of incomplete flowers. They
are surrounded by a kind of
little involucre of modified
leaves, which we might mis-
take for a corolla or perianth.
Each of the stamens which
we find inside the involucre
represents a male flower. Rising from among them on
a short curved stalk is a female flower, consisting of
a three-celled ovary bearing three styles. The Petty
Spurge is a smaller more branched plant, with none of
the golden colour of the Sun Spurge. The umbel has
only three forked branches. The Dwarf Spurge may be
known from both of these by its quite narrow leaves.

Fig. 44. Flower and seed of
Sun Spurge.

The order to which the Spurges belong, which is
named after them *Euphorbiaceae,* is one remarkable
for the powerful medicinal properties of many of its
members. Our own Spurges have a milky juice which
is highly acrid and poisonous.

Euphorbia is the old Greek name for the plant and
its allies, which have long been known on account of
their powerful properties; *Helioscopia* signifies "looking
towards the sun." *Spurge* is the same word as *purge*;
the plant is so called from its medicinal properties.

Couch-grass, ᚢᚱᚩᛁᵯ-ᚠéᚐᚱ (*Triticum repens*).

Couch-grass, Scutch, or Twitch is perhaps the worst
as well as the most abundant of all weeds of arable
land. Its branching perennial underground stems grow
rapidly and vigorously; every broken piece is capable
of forming a new plant; and it is in consequence almost

impossible to exterminate it. It sends up annually leafy shoots a couple of feet in height, the stronger of which bear a spike of flowers reminding one of Wheat, of which indeed it is a close ally. Several other native grasses have a similar creeping habit, but none of them becomes such a pest as the Couchgrass, against which the farmer has to wage perpetual war, by shallow ploughing and harrowing, and gathering the stems thus brought to the surface. The taking of two root crops in succession is effective, as the frequent hoeing that results greatly weakens the weed. Laying down the land to pasture is stated to be a really effective cure. The

Fig. 45. Flower-spike of Couch-grass, slightly reduced. On left a single flower, enlarged.

Couch stems contain much nourishment, and it is recommended to stack them with some lime instead of burning them. In the garden thorough trenching has been found to kill the pest, if the Couch be thrown into the bottom of the trench ; but most gardeners prefer picking it out as they dig.

In the flowers of the grasses such as the Couch we find a floral arrangement very different from any which we have met with hitherto. If we examine the flower-spike of this plant, we shall find that the blossoms are borne in little dense clusters placed on each side of the stem alternately. The flowers, four or five in number in a cluster, consist each of several stamens and a pistil with two styles, enclosed in two little scaly leaves. The whole cluster is enclosed in larger scaly leaves. The outer scales (*glumes*) and the inner ones (*pales*) open to allow the stigma and the delicate stamens with their dangling anthers to project into the air, and close again in a short time when the brief flowering period is over. The flowers of grasses may be studied more easily in many species than in the Couch, for, like some other plants which have rapid and effective vegetative reproduction, flower and seed are not borne to any great extent.

Triticum is the old classical name for the Wheat and its allies, to which the Couch-grass belongs, and is derived from the Latin *tero*, I bruise, because its produce is ground into flour; *repens* signifies "creeping." *Couch, Scutch, Twitch, Quitch,* are forms of an old Anglo-Saxon word from which the word *quake* also comes. The Irish name is ᴠᴘᴏɪᴍ-ꝼᴇᴀᴘ.

BRACKEN, Rᴀɪᴄʜɴᴇᴀch, Rᴀᴄʜen (*Pteris Aquilina*).

Of some forty species of ferns which grow in our islands, the Bracken or Brake is the only one which ever interferes with agricultural operations. In upland districts it is abundant, and frequently invades pasture land in great quantity. It is enabled to do this by its mode of growth. Its stems creep underground; they are thick and strong, and grow rapidly, and send up into the air at intervals the large spreading fronds, sometimes 6 or 8 feet in height, which we know so

well. These shade the ground, and weaken the growth
of useful grasses and so on. As in the case of other
plants of similar growth-form, such as the Creeping
Thistle, the principle on which this weed is to be
attacked consists of starving out the growing underground parts by cutting off the food supply which they
receive from the over-ground parts. Constant scything
or chain-harrowing while the fronds are young, or
beating them down with sticks, all have this effect. A
more complete and drastic remedy is of course the
breaking up of the land.

The Bracken, like all the Ferns, Horsetails, Mosses
and so on, belongs to the great group of the vegetable
kingdom known as "Flowerless Plants." It does not
produce a blossom with stamens and pistil which by
fertilization gives rise to a seed, which is a young plant
capable of growing directly into a plant resembling the
parent. On the under-side of the frond of the Bracken
we shall find carefully protected by the reflexed edge of
each leaflet or *pinnule* an abundance of minute roundish
bodies, brown when ripe, like tiny grains of sand. When
these are mature they open, liberating a great number
of exceedingly minute *spores*, which are "seeds" only
inasmuch as they resemble them in being the resting-
stage of the plant, during which it attains dispersal (in
this case by means of the wind). If one of these spores
lodges in a suitable damp spot, it grows into a little flat
heart-shaped green membrane (the *prothallus*), on the
under-side of which are produced organs corresponding
to the pollen and ovule of the Flowering Plants. When
the female organ or germ-cell is fertilized by contact
with the male organ, it commences to grow, and forth-
with forms a young fern-plant.

The spores of ferns are produced in enormous quantity, and they are very much smaller and very much lighter than the smallest and lightest seeds of Flowering Plants. Nevertheless, in the case of the Bracken it is not seedlings we have to contend with, but invasions due to the rapid growth of the underground stems.

Pteris is the Greek name for Fern, coming from *pteron*, a wing, on account of the wing-like shape of the fronds ; *aquilina* is from Latin *aquila*, an eagle. In Irish ferns are called Raıchneach, older form Rachen (whence Cúıl Raıchın (now Coleraine), "ferny corner," and Raıcheán, now Rahan, in King's County).

INDEX

Heliotrope, Winter 86
Hellebore 25
Hemlock 25, 78
Hemp-nettle, Red 28
  seed 38
Heracleum Sphondylium 76
  seeds 42
Holly 9
Hooked seeds 46
Hordeum murinum 31
  pratense 31
Horse-Chestnut 9, 40
Horse-tails 102

Involucre 79
Ireland, vegetation 8
Iris Pseudacorus 29
Ivy 20, 27

Juncus acutiflorus 52
  effusus 52
  obtusiflorus 52

Killarney, climate 27
Knot-grass 29, 53

Labiatae 25, 93
Lady's-Smock 53
Larch 9
Latin names 61
Lavender 25
Leaves, function 15
Leguminosae 70
Length of life 18
Life 13
Ligule 80
Linaria minor 30
Lolium perenne 31
Loosestrife, Purple 27, 29
Lychnis alba 28
  Githago 29
Lycopsis arvensis 28
Lythrum Salicaria 27, 29

Marigold, Corn 27, 79, 84
Marjoram 25
Matricaria discoidea 30
Meadow-grass, Annual 49
Meadow-sweet 57
Mediterranean vegetation 6
  weeds 7, 29
Migration of weeds 29

Mint 25
Monk's-hood 19, 25
Mosses 102
Mountain Ash 9
Mushroom, spores 40
Mustard, Black 70
  White 69

Natural Orders 61
  Selection 4
Nettle, Common 96
  Roman 96
  Small 96
Nightshade, Enchanter's 46
Nomenclature 61

Oak 9, 10
Oenanthe crocata 78
Onion 19
Orchids, seeds 41
Orobanche minor 24
Osmunda regalis 29
Ovule 65

Papaver Argemone 67
  dubium 67
  hybridum 66
  Rhaeas 66
Parasitic weeds 24
Parsley 78
  Fool's 78
Parsnip 25
  Water 78
Peach 74
Pear 74
Peat bogs 9
Penny Cress 28
Perennials 50
Persicaria 94
Petals 16, 64
Petasites fragrans 86
  vulgaris 86
Phragmites communis 29, 53
Pistil 16
Plantago lanceolata 50
Plant life 1
  names, English 62
    Latin 61
  world 1
Plants, climbing 91
Plum 74
Plume seeds 42

Printed in the United States
By Bookmasters